T0334578

Cambridge Elements ≡

Elements in Theatre, Performance and the Political
edited by
Trish Reid
University of Reading
Liz Tomlin
University of Glasgow

THEATRE REVIVALS FOR THE ANTHROPOCENE

Patrick Lonergan
University of Galway

CAMBRIDGE
UNIVERSITY PRESS

Shaftesbury Road, Cambridge CB2 8EA, United Kingdom

One Liberty Plaza, 20th Floor, New York, NY 10006, USA

477 Williamstown Road, Port Melbourne, VIC 3207, Australia

314–321, 3rd Floor, Plot 3, Splendor Forum, Jasola District Centre, New Delhi – 110025, India

103 Penang Road, #05–06/07, Visioncrest Commercial, Singapore 238467

Cambridge University Press is part of Cambridge University Press & Assessment, a department of the University of Cambridge.

We share the University's mission to contribute to society through the pursuit of education, learning and research at the highest international levels of excellence.

www.cambridge.org
Information on this title: www.cambridge.org/9781009282147

DOI: 10.1017/9781009282185

First published 2023

A catalogue record for this publication is available from the British Library.

ISBN 978-1-009-28214-7 Paperback
ISSN 2753-1244 (online)
ISSN 2753-1236 (print)

Theatre Revivals for the Anthropocene

Elements in Theatre, Performance and the Political

DOI: 10.1017/9781009282185
First published online: August 2023

Patrick Lonergan
University of Galway

Author for correspondence: Patrick Lonergan,
patrick.lonergan@universityofgalway.ie

Abstract: This Element argues that the climate emergency requires a new approach to the study of theatre history – a suggestion that is developed through an analysis of the practice of theatrical revival during the Anthropocene era. Staging old plays in new ways can make visible ecological or environmental features that might have previously gone unnoticed: features which, in some cases, might not have been consciously included by the original authors or makers of a work, but which will be detectable to audiences nevertheless. These links are explored through case studies from the contemporary Irish theatre – including revivals of plays by Shakespeare, Lady Gregory, and Samuel Beckett, as performed by such major Irish companies as Rough Magic, Druid Theatre, and Company SJ. The Element ultimately shows how theatre can contribute to debates about the Anthropocene, and offers new pathways for theatre practice and criticism.

Keywords: ecology, theatre revivals, Anthropocene, Ireland, theatre history

ISBNs: 9781009282147 (PB), 9781009282185 (OC)
ISSNs: 2753-1244 (online), 2753-1236 (print)

Contents

Introduction

Is it meaningful to assert that today's weather was caused, if only in a minor way, by theatre productions from the past? We know that theatre-makers have for centuries burned fossil fuels to illuminate their stages and that, in the contemporary period, many performances involve elaborate lighting and sound effects that consume high levels of energy. There is also a long history of cooperation between theatre and the oil and petrol industries, through corporate sponsorship as well as indirect forms of support. Such impacts are less severe than is the case in such industries as aviation, motoring, or agriculture – but they are not negligible.

But perhaps a more momentous impact has arisen from theatre's capacity to shape attitudes towards the environment, other living beings – and fossil fuels too. It does that in many ways, but primarily by modelling the so-called 'real world' – through scenography, dramaturgy, acting, and other forms of theatrical representation – in ways that have often separated the human subject from (the rest of) nature, a problem exacerbated by the post-Enlightenment tendency in the West to conflate the 'real' with the 'human' while also positing nature and culture as separate categories. In this context, one might think of Shakespeare's *The Tempest*: of how Gonzalo, confronted by the unreality of Prospero's island, delineates the real from the fantastic: 'If in Naples/I should report this now, would they believe me?' he asks, thus performing for the audience the boundaries between fiction and everyday life (3.3.26–7).

That example might point us towards Amitav Ghosh's already-influential book *The Great Derangement* (2016), in which he proposes that one of the causes of the climate crisis is that the Western conception of literary (and thus, one can infer, of theatrical) realism has developed in such a way that when artists attempt to present climate change realistically, the results can seem closer to science fiction or fantasy: 'it is as though in the literary imagination, climate change were akin to extraterrestrials or interplanetary travel', he proposes (p. 7). Through fictions, languages, and other forms of cultural representation, the human species has organized the world in such a way that the climate crisis seems literally unbelievable, Ghosh argues – and that in turn affects readers' and audiences' sense of both urgency and agency in the face of the crisis. Here again *The Tempest* is instructive, its final act presenting Iris and Ceres as human personifications of natural processes – before Prospero admits that they are but 'spirits and/Are melted into air, into thin air':

> And, like the baseless fabric of this vision,
> The cloud-capp'd towers, the gorgeous palaces,
> The solemn temples, the great globe itself,
> Yea, all which it inherit, shall dissolve
> And, like this insubstantial pageant faded,
> Leave not a rack behind. (4.1.148–156)

These lines display possible evidence of an intuition on Shakespeare's part that the use of cultural forms to portray the realities of weather, climate, and the non-human will always prove futile, revealing itself necessarily as artifice, as something insubstantial and baseless: as evidence of a derangement, great or otherwise.

With those ideas in mind, it seems necessary to think about theatre in terms of its propensity not only to mirror but also to structure human attitudes: towards the environment, the climate, non-human living beings, and other features of what has come to be known as the 'natural world' – but which is better understood in terms of Baz Kershaw's definition of 'ecology' as 'the interrelationships of all the organic and no-organic factors of ecosystems, ranging from the smallest and/or simplest to the greatest and/or most complex' (2007, p. 15). As Theresa J. May (2021) points out, stage plays and theatre productions are, and always have been, informed by 'ecological ideologies and implications' that must now be made visible (p. 4). In other words, one of the assumptions underlying my argument is that the modern theatre, especially in the West, has often constituted, promoted, and reinforced 'ecological ideologies and implications'.

Una Chaudhuri (1994) sees this pattern as arising from the fact that theatrical naturalism (and, soon afterwards, realism) emerged in tandem with the spread around the globe of industrialization: theatre, she writes, thus 'hid its complicity with industrialization's animus against nature by proffering a wholly social account of human life. While asserting the deterministic force of environment, naturalism concealed the incompleteness of its definition of environment' (p. 24). There is, then, a need to explore how, why, where, and when theatre has borne responsibility for the ecological crisis – both conceptually (by constituting and reifying attitudes and beliefs) and materially (by engaging in practices that are destructive of the environment). But this Element also seeks to substantiate the assumption that theatre has played a positive role in raising ecological awareness: that it has offered alternative modes of engaging with the ecological, that it has developed and used environmentally responsible forms of artistic practice, and that it has encouraged audiences to take responsibility for their own actions. Those characteristics may also be identified in the past – perhaps may even be retrieved from the past, to be applied anew in the present.

My intention is to discuss these matters by exploring the theme of theatrical revival – and, by doing so, to find common ground between the conceptual and material perspectives described above. For the present purposes, I define 'revival' as the re-staging of dramatic performances in contexts and/or locations that were not necessarily imagined when the work was composed and premiered. Revivals can, I suggest, make visible ecological or environmental features that might previously have gone unnoticed: features which, in some cases, might not have been consciously intended by the original authors or makers of a theatrical performance, but which will nevertheless be detectable and meaningful to audiences in later eras. Theatrical revival also necessarily performs a dialogic relationship between the present and the past: it is theatre historiography in action. Revival requires theatre-makers to understand and form attitudes to their history – but it also obliges them to give physical expression to their interpretation of that history, to render the conceptual in material form. Revival therefore necessarily re-enacts 'ecological ideologies and implications' from the past and in so doing can inspire agency in the present and, perhaps, hope for the future; it can therefore be seen as an example of what Kershaw terms 'performance ecology': 'a discipline . . . that aims to refigure the relationship between "culture" and "nature" that all humans inevitably inherit' (p. 15).

I aim to develop these claims by exploring case studies from the contemporary Irish theatre: revivals of plays by Beckett, Shakespeare, and Lady Augusta Gregory, all performed between 2018 and 2021. Shakespeare and Beckett are, of course, produced widely and often internationally – but, even so, my decision to prioritize a single national tradition requires some explanation. It arises from an interest in one of the methodological problems associated with the study of cultural ecology: namely, the difficulty of tracking the relationship between human agency and a set of phenomena that are vast in relation to both space and time, and which include climate, ecological interconnection, geology, and other 'hyperobjects' (to use a term coined by Timothy Morton and developed in detail in that author's eponymous 2013 book) that lie beyond the perceptual powers of the individual human, whether scholar, audience-member, or theatre-maker. Therein lies one of the risks that studying theatre and ecology entails: scholars must find meaning in case studies that, if too narrowly constituted, will be inconsequential but, if too broad, will lead only to generalizations or false claims of universality. I want to explore whether it is possible to find space between the irrelevantly small and the unknowably vast by discussing the impact of single productions that have emerged from what Morton terms 'monstrously long' timespans: productions that cut across national, linguistic, and formal boundaries: entities that are often so 'massively distributed that we can't directly grasp them empirically', as Morton puts it (2016, p. 11).

I also see the Irish example as being valuable because it is distinctive – and because I consider that the country has experienced a diverse range of phenomena that allow for comparative perspectives to emerge, both to Western and non-Western traditions. Ireland has, for example, experienced both colonialization and globalization: it was a victim of the former for almost 800 years but has been a beneficiary of the latter since the 1990s. In response to colonization by England, Ireland developed a form of cultural nationalism that continues to define the operation of its theatre in the present; that national dramatic tradition asserted links between identity, landscape, language, and the natural world – developing a model that was inspired by European romanticism, but which would also inspire postcolonial theatres in Africa, South America, and India. Ireland has strong cultural ties to Britain and the United States – two countries that bear great responsibility for environmental destruction – but, unlike those countries, Ireland has a relatively limited store of fossil fuels, did not experience large-scale industrialization until the late twentieth century, and has made a comparatively small contribution to the pollution of the planet. Ireland is part of the English-speaking world, and has benefited economically and culturally as a result – but it also has an indigenous language (*Gaeilge*) that conveys an interconnection between humans and landscape, weather, and the rest of the biosphere; that indigenous culture was supplanted (but not eradicated) by English imperialism – giving Irish artists a dual perspective that has long been described as one of the defining features of its culture. Finally, and perhaps most pertinently for the argument outlined in this study, Ireland's modern culture has been influenced by the experience of a traumatic ecological collapse: the Great Irish Famine of the mid-nineteenth century, which caused the population of the island to fall from eight million in 1845 to roughly half that number in 1900. For these and many other reasons, a focus on Ireland can allow for the testing of ideas that might also be applicable in other geographical and historical settings, partly because Ireland has had experiences that cut across so many international, chronological, and ecological boundaries.

I have also begun with the proposition that the history of theatre, especially in the West, is interconnected with environmental histories – and specifically with our planet's descent into a new epoch that has come to be known, not uncontroversially, as the 'Anthropocene'. That is a contested term that connotes an incontestable fact: that the human species is facing the simultaneous and interrelated crises of climate change, ecosystem collapse, mass extinction, deforestation, and ocean acidification – not to mention the failure of those in positions of power to meaningfully address the role of human decision-making in those crises.

In these pages, I am attempting to consider how a consciousness of environmental exploitation, neglect, and destruction can be tracked in and through theatre histories, and through the material reflection of theatre historiography that the practice of revival requires. But there is a need to address the suitability of 'the Anthropocene' as a term for achieving that goal. The word's origins lie in Geology, and with a proposal that the impact of human activity upon the planet has become so strong as to require the designation of a new geological epoch, one that would follow the interglacial period of relative warmth and stability that began approximately 11,000 years ago, and which is known as the Holocene. That proposal was made by Paul Crutzen at an event in the year 2000, and subsequently developed in an article published in the year 2002. An Anthropocene Working Group (AWG) continues (at the time of this writing) to debate whether the term should be formally adopted as a Geological time scale, but the word has long since escaped the boundaries of its discipline of origin so that, as the AWG (2022) itself states, it 'has developed a range of meanings among vastly different scholarly communities'. That gives rise to the problem that there are many different 'Anthropocenes', and that the term may therefore occlude disciplinary assumptions rather than allowing for cross-disciplinary understanding. It will therefore be necessary to define what the Anthropocene might mean for theatre studies, a task that I hope this Element will advance (though it does not claim to offer either the first or the final word on the subject).

More seriously, many scholars have pointed out that one of the problems with the use of the Anthropocene term is that it appears to attribute agency in relation to the climate and biodiversity crises to the species (*Anthropos*) when the primary responsibility lies with a small minority of humans (mostly inhabitants of Western industrialized nations). Accordingly, a wide range of alternative terms have been proposed. The most common of these is the 'capitalocene', a designation that aims to make explicit the link between capitalism and environmental destruction (see Moore, 2016; Davis, et al. 2019; for theatre studies, see Gillen, 2018 and Arons, 2020). That word has also sometimes been criticized for overlooking the extent to which other modes of societal organization have also entailed environmental destruction (the link between Soviet communism and the 1986 Chernobyl disaster is sometimes mentioned as a case in point), but it has the virtue of making clear how capitalism is undoubtedly a major cause of the ecological crisis, as well as being one of the primary barriers to addressing it. Also proposed are such terms as Haraway's 'Chthulucene', though she also uses the terms Anthropocene and Capitalocene as appropriate (2016). Then there are such words as 'phallocene', 'thanatocene', 'plantationocene', and others that are outlined by Bonneuil and Fressoz (2016). Also of great significance is the feminist critique of the Anthropocene paradigm, as explored especially by Stacy Alaimo (2017; see also Stevens, Tait, and Varney, 2017). Scholars such as Rob Nixon (2011) and Kathryn Yusoff (2018) have also argued that the history of the Anthropocene must be linked to the histories of colonialism, capitalism, and so-called modernization – phenomena that originated mostly in European contexts. And within theatre studies some scholars, such as Wendy Arons (2020) and Aleriza Fakhrkonandeh (2021), have offered strong arguments for the unsuitability of the Anthropocene paradigm altogether.

This Element does not aim specifically to defend the use of the Anthropocene term over all available alternatives, but instead to accept that the word exists and has currency both within and beyond the sciences. I am also influenced by Vicky Angelaki's exploration of how the term requires notice within theatre studies because its existence is indicative of a growing environmental awareness (2019, p. 7), and am sympathetic too to the suggestion that, for all its flaws, the word has entered public discourse in a way that allows members of the public to discuss meaningfully the role of human activity in environmental change and collapse. It has, for example, been included in the *Oxford English Dictionary*, has been used in the title of popular albums by Grimes and Renée Fleming, and has begun to appear in the names of university courses and research centres around the world. Following the lead of Varney (2022) and others, I also consider that it is possible to view the Anthropocene as an epistemic and investigative framework that can attend to such issues as capitalism, patriarchy, colonialism, and other phenomena that, some fear, risk being obscured by the use of the term *Anthropos*. I also hope to demonstrate the validity of the Marxist scholar McKenzie Wark's observation that a 'theory for the Anthropocene can be about other things besides the melancholy paralysis that its contemplation too often produces'. He elaborates on that claim by highlighting the need to get 'to work on the kinds of knowledge practices that are useful in a particular domain' (in the present case, the domain of theatre scholarship and practice) rather than becoming debilitated by the larger problem (2015).

My goal, then, is to 'get to work' – to show that, since the word exists anyway, it must be properly nuanced so as to make the Western, capitalist, and patriarchal roots of environmental crisis more visible – or, as Peter Sutoris puts its more succinctly, 'Instead of getting rid of this term, let's decolonise it' (2021). Yes, it is necessary, following Haraway, to be aware that 'the Anthropocene obtained purchase in popular and scientific discourse in the context of ubiquitous urgent efforts to find ways of talking about, theorizing, modeling, and managing a Big Thing called Globalization' (p. 45). But the adoption of the Anthropocene term need not require the abandonment of the more nuanced and specialized terms mentioned above: David Farrier, to give just one example, successfully uses several of them in his *Anthropocene Poetics* (2019). Having said that, I also accept that there is a need to attend fully to the concerns and objections of the scholars mentioned above. Ultimately, therefore, I am seeking to re-apply a call made by Alan Read in *Theatre and Everyday Life* (1995): 'what is needed', he writes, is 'not the ignorance of nature but more acute definition of the links between political, ethical and creative progress and living within nature, which inevitably is a transformation of nature' (p. 140). Theatre – and theatre criticism – have the potential and the responsibility to define those links.

Because of that impulse to place the 'Anthropocene' term in a broader cultural context, I am also interested in considering how theatre can contribute to one of the primary debates about the Anthropocene hypothesis, which concerns the determination of an appropriate starting point for it. Most scientists suggest that it began with the detonation of the first nuclear bombs in 1945 (initiating a period known as the 'Great Acceleration', which is discussed in the first section); others argue that it began in the 1760s with the industrial revolution, still others that it began in 1610, and a small number that it began thousands of years ago, with the invention of agriculture. Within the field of Geology, that debate is being conducted in relation to such considerations as the chemical composition of the atmosphere and the presence of pollutants in rock strata – and it is likely that the post-1945 era will be selected as the 'official' starting point. But, as Simon L. Lewis and Mark A. Maslin wrote in an influential article for *Nature* (2015), the outcome of that decision will have ethical as well as scientific consequences: 'The choice of 1610 [...] as the beginning of the Anthropocene would probably affect the perception of human actions on the environment,' they note. 'The Orbis spike [which occurred in that year] implies that colonialism, global trade, and coal brought about the Anthropocene. The event or date chosen as the inception of the Anthropocene will affect the stories people construct about the ongoing development of human societies' (p. 180). Kathryn Yusoff expresses the same idea with greater force and concision: 'Origins,' she writes, 'are another word for an account of agency or a trajectory of power' (p. 25).

The 'Orbis Spike' is a phrase developed by Lewis and Maslin to describe the fact that the year 1610 marked a low-point in the concentration of CO_2 in the atmosphere (the 'spike' denoted by the phrase is visible in graphs that track that presence over several centuries, and is based on measurements of Antarctic ice cores). That low-point was almost certainly caused by the colonization of the Americas and the subsequent genocide of indigenous populations there. Explained simply but, I hope, not simplistically, the claim is as follows: the arrival of Europeans and their diseases after 1492 caused tens of millions of Americans to die; forest regrowth occurred on the land that those people had been farming; that new vegetation sequestered large amounts of carbon, allowing more heat to radiate back to space and thus causing global temperatures to fall – a process that reached a 'spike' in 1610 before rebounding as land in the Americas returned to agricultural use. The term 'orbis' is derived from one of the Latin words for world, intending to capture the fact of human planetary interconnection that was initiated with the Columban 'discovery' of the Americas. So, as Lewis and Maslin point out, beginning the Anthropocene in 1610 would have moral implications, since it would inextricably link the ecological crisis of the present to such causes as colonization, the genocide of indigenous populations in America, and the emergence of the trans-Atlantic slave trade – all of which have their origins in the early modern period, and in Europe. It would also make clear how human decision-making can have consequences that long outlast the duration of a single human life.

For theatre scholars, mention of the year 1610 might inspire further thoughts of Shakespeare's *The Tempest* – a play that was first performed in 1611 but which was probably written during the year before, having (it is speculated) been inspired by a written account of a shipwreck in Bermuda from 1609 (as discussed by Mentz, 2015, pp. 54–6). *The Tempest* could also be read in relation to Lewis and Maslin's argument about the Anthropocene: it is a story that 'people construct about the ongoing development of human societies' and it certainly has much to say about 'colonialism [and] global trade': Mentz sees the play as existing firmly within the Anthropocene, while acknowledging the problems with that word: 'climate change may be our fault', he writes (the 'our' referring to people living today in the West), 'but it is not only *our* world' (emphasis added; 2015, p. xvi). Plays such as *The Tempest* can be used to emphasize that the collective *Anthropos* in 'Anthropocene' refers not to universal human responsibility for environmental destruction, but rather to the necessarily universal character of the extinction of the human species.

Of course, Shakespeare knew nothing about the physics of climate change. But what might it mean to revive *The Tempest* in the present if audiences started to think of it as one of the first dramas of the Anthropocene? Whether arising from coincidence, correlation, or causality, it must be acknowledged that there is a chronological relationship between modern theatre and environmental history. Wolfgang Behringer (2010) has found evidence of new ways of seeing the world in literature in several sources from early modern Europe, finding examples not only in the plays of Shakespeare but also in Cervantes, Andreas Gryphias, and elsewhere (p. 144). Early modern European drama emerged in the sixteenth century; it therefore seems worth exploring the fact that the Orbis Spike coincides with the appearance of European plays, including *The Tempest*, that offered audiences different methods of understanding human interaction with planetary forces. In such a context, how might one interpret Thomas Dekker's decision to dedicate his *Satiromastix* (1601) 'to the world'? What to make of Ben Jonson's *The Magnetic Lady* (1632), and its use of planetary magnetism as a metaphor for the relationships between his characters? Are new approaches possible for reading and staging Lope De Vega's *La Dama Boba* (1613) or Calderón's *El Gran Teatro del Mundo* (c. 1634)?

It might also be possible to track how the Industrial Revolution coincided with – and allowed for – the introduction of new forms of stage technology, much of it dependent upon the consumption of fossil fuels. In the early 1800s, for example, London's Lyceum Theatre became one of the first venues to use gas and oil for stage lighting; and electrical lighting began to be used at the Paris Opéra as early as 1849. And in the contemporary era, the 'Great Acceleration' period has coincided with a diversification of theatre practices in ways that have often mapped on to broader societal trends. Theatre has become both more experimental (partly because of increased state funding in the post-war period) while simultaneously becoming more expansive, as shown by the growth, from the 1970s onwards, of 'mega-musicals' such as *Les Misérables*, which require huge casts, and which seek to generate huge profits – mirroring broader trends in late capitalism towards nichification and massification.

The preceding two paragraphs offer a very brief sketch of a very broad field of knowledge, but in doing so they seek to illustrate that it is at least *possible* to use theatre historiography to test the idea that, if the Anthropocene *did* begin with the Orbis Spike – and thus with the development of early modern theatre – then it should be possible to find evidence of that development from the study of plays and performances from the 1500s to the present. By considering how the origins of the Anthropocene may be identified not just in rock strata and ice cores, but also in material cultural forms such as stage plays (including scripts, set designs, lighting designs, and so on), theatre scholars might be in a position to argue that our research can propose answers to questions that have proved inconclusive in such disciplines as Geology, Chemistry, and Marine Science.

Ecocritical and green approaches to theatre studies have been in use since at least the 1990s, and have been explored through monographs, journal special issues, articles, conferences, and symposia. Some of those publications predate the coinage of the Anthropocene term; others decline altogether to use that word – but in general, this scholarship has been invested in exploring how theatre can effect change in the present, through sustainability, activism, policy development, and other forms of transformative activity. The most extensive scholarship has involved the investigation of the relationship between contemporary theatre practice and ecology, often emphasizing the power of theatre to advocate for environmental justice; key examples include the scholarship of Vicky Angelaki (2019), Una Chaudhuri (1994, 1997, and 2013 especially), Baz Kershaw (2007), and Carl Lavery (2019), as well as Giannichi and Stewart's 2005 collection *Performing Nature*. And, in the context of Irish theatre, Lisa Fitzgerald's *Re-Place* (2017) offers a pioneering investigation of the interrelationship between environment, site-specific performance, and materiality. A second strand explores the concept of ecodramaturgy, a term (discussed in more detail in the next section) that involves the reading and/or reinterpretation of dramatic texts to retrieve, reveal, or impose ecological perspectives – as seen in the scholarship of May (2005, 2021), Arons (2020), May and Arons (2012), and Woynarski (2020). Relatedly, a third strand applies ecocritical tools to the interpretation of dramatic texts, drawing on theoretical methods that have also been applied to re-reading fiction, poetry, and other forms of art; such scholarship often prioritizes single writers or periods (ecological approaches to Shakespeare's drama are particularly advanced – see Egan, 2006 and 2015; Bruckner and Brayton, 2011; Martin, 2015; O'Malley, 2020). The growing field of ecosceneography is also important for its analysis of the material impacts of theatre design and its consideration of the sustainability of theatre practice (as discussed by Julie Hudson, 2020 and Tanja Beer, 2021). There have also been noteworthy studies that explore older dramas, exploring how such plays might be, or have been, revived in ecocritical contexts, as shown in O'Malley's analysis of outdoor Shakespeare performances (2020) and by Cless's *Ecology and Environment in European Drama* (2010), which considers contemporary practice-based approaches to plays by Aristophanes, Marlowe, Shakespeare, Giraudoux, Brecht, and Chekhov. There is also a growing body of work that explores the intersections between theatre and theory – including the use by theorists such as Bruno Latour (2017, pp. 28–33) of theatre as a mode of theoretical enquiry, as well as the application of theoretical concepts to the ecological analysis of plays and performance by theatre scholars. Timothy Morton's work has been especially influential, as shown by Aston's exploration of dark ecology in Churchill (2015), Prateek's exploration (2020) of the same concept in relation to Ibsen's *Peer Gynt*, and Ahmadi's discussion of hyperobjects in relation to the drama of Andrew Bovell (2015). In what follows I will myself draw sometimes from Morton's ideas.

Of special importance is a growing number of publications that consider the theme of indigeneity and its relationship to the Anthropocene. Scholarship on that theme to date has often focussed on the indigenous cultures of north America and Australia. This includes studies such as those by Schafer (2003), Simmons (2019), Varney (2022) and Whyte (2017) – but of particular value are the ideas of Helen Gilbert (2013b, 2013a, 2014, 2019, 2020), which have provided methodological and analytical models for understanding how ecologically insightful dramas and productions have been created in contrast with, in isolation from, and/or in opposition to such intellectual formations as modernity, capitalism, or imperialism. 'At its widest scale', she writes.

> indigeneity now operates simultaneously as a portmanteau category establishing community among different peoples with distinct histories and geographies .... and a heuristic framework for thinking about that commonality in relation to origins, affiliations, cultural genealogies and place-based connections. In turn, this framework, in conjunction with on-the ground activism it underpins, has begun to exert pressure on international relations in subtle ways. (2013b, p. 174)

Those characteristics demonstrate the power of indigenous performance practice to produce new forms of knowledge and to work towards the decolonization of the Anthropocene concept (a point elaborated upon in the edited collection *Ecocritical Concerns and the Australian Continent*, to which Gilbert contributes, 2020). However, they also demonstrate the need for a nuanced approach to indigeneity – especially in an Irish context. Ireland is, after all, unquestionably the home of an indigenous culture that has a lineage that intersects with colonial and environmental histories. Yet Irish people, acting independently and/or as members of religious orders and imperial armies, have also been responsible for the oppression of indigenous cultures, including in Canada, the United States, and Australia. There is a need for the scholarship to be able to accommodate both elements of that history. That in turn might also explain my desire, following Yusoff and others, to ensure that the Anthropocene concept be embedded in early modern histories of colonialism.

To date, much (though certainly not all) of the scholarship on the Anthropocene has been dominated by the 'Great Acceleration' period. There is, for example, a special issue of the *Nordic Theatre Studies* journal on *Theatre and the Anthropocene* edited by Steve Wilmer and Karen Vedel (2020), a publication that merits praise for including diverse forms of theatre practice from many countries and languages – but which mostly explores performances that have been produced since the Second World War, such as the politics of eco-performance in late 1970s Sweden, or a collaboration between Asger Jorn and Guy Debord in the 1950s. Likewise, Ahmadi's *Towards an Ecocritical Theatre: Playing the Anthropocene* (2022) is a theoretically rich dramaturgical analysis of contemporary Anglophone plays. There have also been several articles and chapters that relate the Anthropocene hypothesis to contemporary dramas such as Caryl Churchill's *Far Away* (2000), Chantal Bilodeau's *Sila* (2012), Lucy Kirkwood's *The Children* (2016), and Ella Hickson's *Oil* (2016) – but, by staying in the contemporary period, such scholarship tends to display more interest in reflecting awareness of the Anthropocene in the present rather than attending to its historical origins (see Angelaki, 2019; Lonergan, 2020, 2022; Balestrini, 2020). None of those remarks is intended as a negative criticism; rather, I make them to highlight how this study is trying to do other things.

An emphasis on the contemporary period is apparent in other literary and cultural fields of scholarship that explicitly name the Anthropocene as a field of enquiry. *A Cambridge Companion to Literature and the Anthropocene* (2021) defines an emerging canon of Anthropocene texts, for example – with most of its case studies from the late twentieth and early twenty-first centuries, as shown in its discussion of novelists (Richard Powers, Karen Tei Yamashita, Michel Houellebecq), poets (Charles Olsen), film-makers (Gore's *Inconvenient Truth*), writers of non-fiction (Ghosh, Robert MacFarlane), and dramatists (Hickson's *Oil* again features). More generally, there have also been articles and chapters that explore the Anthropocene in relation to other contemporary writings, including novels by Margaret Atwood, Jeanette Winterson, and Ian McEwan; and poetry by Alice Oswald, Ted Hughes, and Moya Cannon. There has also been extensive scholarship on 'cli-fi' – the use of science fiction to explore the Anthropocene, whether explicitly (as in the novels of Kim Stanley Robinson) or *avant la lettre*, as in the environmental fictions of Octavia E Butler or Ursula LeGuin (see Streeby, 2018).

How, then, can these critical approaches to the Anthropocene be applied to theatre history and to the related practice of theatre revival? To answer this question, it may be helpful to consider broader scholarly approaches to the problem of measuring 'deep time' – that is, the time scale of geological events, which is usually so vast as to be imperceptible if not incomprehensible to humans – despite the fact that, as I have discussed above, human responsibility for environmental destruction can outlast a single human life and may continue to have an impact on the planet long after the human species has itself become extinct. Dipesh Chakrabartry has written about how 'as humans we presently live in two different kinds of "now time" simultaneously: in our own awareness of ourselves, the "now" of human history has become entangled with the long "now" of geological and biological timescales, something that has never happened before in the history of humanity' (p. 7). Theatre studies must therefore be conceived of as operating in relation to both human and non-human timescales, especially in the context of the practice of reviving old plays in the present. Cless explains this need by drawing a parallel between ecology and theatre practice. Ecology, he writes, can be 'extremely broad in perspective, such as [when it considers] the whole biosphere, or highly specific, focussing even on one organism and its immediate environment', which he links by analogy with a 'long view of European drama's relation to nature and [the need for a] close, deep investigation of particular plays and playwrights' (2010, pp. 4–5). This Element aims to draw out further the importance of that analogy.

Theatrical revival provides a context for research and analysis in relation to both the specific 'present' that is the moment of a single live performance, and the vast 'now' that is bridged when a play or performance is re-staged after a gap of decades or even centuries. Revivals can, in short, concretize the varying and sometimes divergent perceptions of the experience of time and its relationship with human agency, and – as the examples in this Element seek to illustrate – the completion of that task can have value, both within and beyond the theatre.

Such an approach can also offer new ways of thinking about the disciplinary boundaries of theatre studies, which often involve the assumption of spatial and/or political limits that are concurrent with the timespan and perspectives of a single human life. Research periods, for example, are sometimes defined directly in relation to human lifetimes, as in the description of Shakespearean theatre as occurring in the 'Elizabethan' and 'Jacobean' periods, or indirectly, as in the example of German Romanticism (c. 1790–1850), and the Irish Dramatic Revival (c. 1890–1940), among many other examples that present literary or cultural activity within a roughly 50- to 80-year frame. Even the designation of the present period as 'contemporary' implies a human-centred perspective on the causes and effects of theatre practice. Yet one of the primary problems of the Anthropocene is that it requires humans to take responsibility for how their actions will affect generations not yet born. By tracking the present-day impact of choices made during the Elizabethan period, the *sturm und drang* era, and so on, scholars can better understand interconnections between past and present, while also forming new models for measuring how decision-making now might affect the planet in the future.

One way to adopt that broader perspective is to draw out the implications for theatre of Andreas Malm's discussion of fossil capital (2016). Fossil capitalism can be defined as the concept that cultural norms have developed that render the link between capitalism and fossil fuels appear necessary and 'normal'. This means that, for theatre-makers, the use or the representation of fossil fuels often appeared inevitable rather than as an outcome from a set of choices. Correspondingly, it also means that stage plays and other forms of theatrical representation (such as scenography) have presented the use of fossil fuels as normal or, again, as inevitable. For example, the most frequently used prop at the Abbey Theatre during the twentieth century was not a pint glass, a statue of the Sacred Heart, or a weapon – but a sod of turf (that is, a fossil fuel): a relatively innocuous object that appears in more than half of the Irish plays staged at the national theatre between 1904 and 2000, simultaneously in plain sight and invisible.[1]

Methodologically, such an approach requires aligning theatre histories to carbon histories, and tracking where there are divergences and overlaps. For example, in *Fossil Capital*, Malm describes petrol-burning cars as 'invisible missiles aimed at the future', reasoning that their emissions 'will have their greatest [climate-altering] impact on generations not yet born' (2016, p. 12). If one accepts the accuracy of that statement, then it becomes vital to consider whether theatre, too, can be considered a kind of missile directed at the audience's descendants – given that theatre too has burned fossil fuels, shaped attitudes, and otherwise had a causal role in the creation of this planet's present

---

[1] This determination was made through consultation of the Abbey Theatre Digital Archive at University of Galway.

and future. Again, theatrical revival can reveal those 'missiles' being primed and launched in the past and, by doing so, might draw awareness to decisions being made now. This Element will emphasize the value of Malm's approach to history by choosing as case studies a group of plays that come from different eras in the evolution of fossil capital: the early modern period (Shakespeare), the early twentieth century (Gregory), and the beginning of the 'Great Acceleration' period (Beckett).

Another way to understand the metaphor of theatre as a 'missile to the future' might be to draw on Mark Bould's 'Anthropocene Unconscious' hypothesis, which aims to develop methods for tracking evidence of how the 'Anthropocene and especially anthropogenic climate destabilization manifests in texts that are not overtly about climate change; that is, unconsciously' (2021, p. 2). Bould's approach offers a valuable solution to a problem posed by scholars such as Fressoz and Harraway – which is how cultural forms (including theatre) may have allowed Western societies to *suppress* knowledge of humans' environmental impact. Writing in *L'apocalypse joyeuese* (2012), Fressoz argues that the 'discovery' of the Anthropocene in the early twenty-first century was possible only through a process of deliberate suppression: modernity, he writes, is 'not the story of awareness but a history of the scientific and political production of a kind of modern unconsciousness' (*'une certaine inconscience modernisatrice'*) (2012, p. 11, my translation). In other words, the human species, especially in Europe, has been ignoring the evidence of its capacity for planetary destruction for hundreds of years. More recently, in *Les Révoltes du ciel* (2020), a book co-written with Fabien Locher, Fressoz shows how the history of climate change has its roots in the fifteenth century, proposing a re-reading of the past that will make that history visible and comprehensible to all. Those perspectives chime with the views of Donna Haraway who, in *Staying with the Trouble* (2016), argues that the period since 1990 ought not to be called the Anthropocene but 'the dithering', on the basis that the first IPCC report (published in that year) had rendered undeniable the link between global warming and human activity (p. 102). An exploration of theatre history can speedily demonstrate that the 'Anthropocene' is a new word for a concept that has been intuited if not understood since at the last the 1800s.

By attending to symbolism, images, and other non-literal forms of representation, Bould finds evidence of attitudes to the Anthropocene in film, fiction, poetry and more – showing how it can be possible to find evidence of *'une certaine inconscience modernisatrice'* or 'dithering' in literary texts and performances. There are risks involved in this approach: to propose that ecological material appears 'unconsciously' in a dramatic work might invite the making of claims that can never be validated. Here again the alignment with broader fossil histories is essential: by showing areas of coincidence, equivalence, and overlap between theatre and broader social forces, it becomes possible to argue for the presence of new motivations and insights. Bould's use of the term 'unconscious' is not intended to refer to the identifiable or diagnosable psychological entity – but nor is it exactly intended as a metaphor. Rather, it is evidence of a way of seeing the world, a perspective that can be reconstructed and argued for as a form of knowledge.

One frequently used example of this phenomenon is, as scholars of Shakespeare and ecology frequently point out, to be found in *A Midsummer Night's Dream*, when Titania describes how her arguments with Oberon have had climate-altering impacts. 'With thy brawls thou hast disturb'd our sport', she says:

> Therefore the winds, piping to us in vain,
> As in revenge, have suck'd up from the sea
> Contagious fogs; which falling in the land
> Have every pelting river made so proud
> That they have overborne their continents [. . .]
> And through this distemperature we see
> The seasons alter. (2.1. 88–107)

Shakespeare may have been writing about fairies, but he was also showing an understanding (or, more accurately, another possible intuition) that human actions can have planetary repercussions. This speech conveys a sense of how conflict causes the disruption of boundaries: even the verse seems determined to break through its limits, with the words 'sea', 'land' and 'proud' spilling immediately and without a relieving punctuation mark into the next lines. It is helpful to think about what this writing requires of the body of the actor, to imagine the breathing required to deliver, without a break, the eighteen words from 'which falling' to 'their continents'. This short speech thus offers a wonderful blurring of usually separate categories, dramatizing an interconnection between the printed word, the actor's breath, the audience's imagination, and the planet's climactic systems. It also illustrates the validity of Tom McFaul's suggestion that Shakespeare's plays encourage humans to 'accept their relationship to the natural world' (p. 193). Perhaps, then, one can think of this speech as another kind of missile to the future: an unconscious manifestation of a future that was beginning to form when Shakespeare wrote his plays.

Another way of changing the chronological perspective of theatre studies might be to draw on the idea of 'future fossils'. That concept has become increasingly common in museum curation and visual arts: there is a Museum for Future Fossils project that presents material online; similarly, the Museum

De Lakenhal in Holland exhibited a 'Future Fossils' exhibition by Walter van der Velden and Aart Kuipers. There is also a Future Fossils podcast by Michael Garfield, which combines art, philosophy, and palaeontology; and a scholarly formulation of the concept can be found in Farrier's *Footprints* (2021). While each uses the term slightly differently, all have in common a desire to capture how objects created today will have an environmental impact hundreds or thousands of years from now. To give a frequently used example, a Styrofoam container for an interval snack during a theatre performance can exist for hundreds of years, but the food itself might take less than five minutes to consume: thus, the use-value of the object is out of synch with its long-term environmental presence (see Farrier, 2019, pp. 55–88). By seeing such objects as 'future fossils' rather than in relation to their short-term utility to the individual human consumer, one might resist the tendency to suppress or ignore knowledge of how human actions will affect future beings.

Within theatre practice, the concept of 'future fossils' can be applied to the use of materials such as plastics (which are a major feature of costume, set and lighting design in live performance). But the concept also represents a way of seeing all activities in relation to their long-term geological, social, and environmental impact. Identifying contemporary objects, actions, and ideas as potential 'future fossils' (whether literally or metaphorically) allows, encourages, and obliges scholars and theatre-makers to imagine ways of making the 'Anthropocene Unconscious' manifest.

These are the theoretical ideas that this Element is founded upon. It begins with a consideration of an outdoor Irish production of Beckett, arguing that an ecodramaturgical approach to staging his plays can intersect with other political and aesthetic objectives. This argument is explored through an investigation of a revival of *Happy Days* (1961), which was translated into Irish by Micheál Ó Conghaile and performed on the Aran Island of Inis Oírr by Sarah Jane Scaife's Company SJ in 2021. Beckett's plays are increasingly being seen as sources of understanding of the climate crisis, but also significant is that the translation of *Happy Days* into the Irish language allowed for the retrieval of knowledge from the indigenous culture: knowledge that was latent in Beckett's original and which was, in Scaife's revival, brought to light.

Section 2 considers a revival of Shakespeare's *Midsummer Night's Dream*, directed by Lynne Parker for her own company Rough Magic in 2018. The discussion considers how her staging encouraged audiences to think about the effects of human decision-making over several centuries. A key feature of that section is to identify the fact that 'Orbis Spike' period – and the emergence of early modern English drama – coincides with the colonization of Ireland. The confluence of these strands in Rough Magic's staging allows for a consideration of the overlaps between colonialism, modern theatre, and climate crisis – providing practical ways of thinking about Kathryn Yusoff's analysis of the relationship between the Anthropocene and early modern imperialism.

The final section considers how Druid Theatre's outdoor staging of the plays of Lady Gregory in 2020 presented those dramas as historical artefacts (they are of their time – that is, the early twentieth century) while also showing how they were revitalized by being performed in new contexts. Lady Gregory's plays contain a store of knowledge about Irish wildlife, agriculture, and food production; Druid's production of her plays in the outdoor setting of Coole Park (the west of Ireland estate that Gregory occupied until her death in 1932) exposed that knowledge. By doing so, the production explores the concept of 'ecognosis' – a word that refers not only to the acquisition of ecological knowledge but also to the development of modes of thinking that are better attuned to ecological concerns.

1 Ecodramaturgies: Beckett's *Happy Days/Laethanta Sona*

The dramatic *oeuvre* of Samuel Beckett can be seen as a source of exemplary plays for the Great Acceleration period. That era was initiated by the detonation of the first atomic bombs in 1945, an occasion that (as mentioned in the introduction) is seen by many scientists as representing the starting point of the Anthropocene. It is a phrase that aims to capture the intensification from the 1950s onwards of many processes that had hitherto been increasing more slowly and thus more sustainably: not just the elevation of carbon dioxide and methane levels in the atmosphere, but population growth, deforestation, fertilizer usage, biosphere degradation, and many other indicators. When graphed across time, those numbers proceed horizontally for centuries before ascending sharply, providing a vertiginously chilling image of (some of) the human species' worsening impact upon the Earth. By visualizing environmental history so effectively, those graphs have themselves 'become an iconic symbol of the Anthropocene, and have been reprinted many times and in many different academic and cultural fora and media', as Will Steffen et al. point out (2015, p. 18). Those authors also mention that a version of the graphs 'even appeared in Dan Brown's novel *Inferno*' in 2013 – providing a thought-provoking example of how the Anthropocene paradigm, despite its limitations, can enable the transmission of meaningful ecological information to large numbers of people, in this case through a novel that sold 250,000 copies in the UK alone during its first week of sale (see Flood, 2013).

What distinguishes the 'Great Acceleration' from other periods associated with the Anthropocene is not just the speed of change, however – but also the fact of human awareness of the species' *responsibility* for that change. We now know that humans have the capacity to destroy life on the planet, whether through nuclear war, anthropogenic climate change, the misuse of land and the oceans, or many other causes. That awareness produces a sensation that might in its own way be conducive of vertigo: a feeling of severe disjunction between, on the one hand, the helplessness of the lone individual (who is statistically if not actually blameless for these crises) and, on the other, the collective (though not universal) responsibility of the species for planetary upheaval. That mismatch produces a kind of ecological version of Heidegger's *Geworfenheit*, a sense of being 'thrown' into an existence that one did not choose but is nevertheless responsible for. *Waiting for Godot* (1955), *Endgame* (1957), *Happy Days*, and other Beckett plays are not in any direct way deliberately or consciously *about* those themes, but they are of their time in placing awareness and the need for action into tension with each other, asking what must be done when one is 'thrown' into a world in which there is 'nothing to be done'. To revive such dramas in the Anthropocene era might therefore involve allowing audiences to receive Beckett's plays in ways that he might not have intended but which should nevertheless be seen as legitimate.

Perhaps because of such features, Beckett has attracted extensive attention from ecocritical scholars, over several decades. Writing in 1994, Una Chaudhuri described *Endgame*, together with Ibsen's *Enemy of the People* (1882), as essential examples of the 'largely negative ecological vision' that had permeated the twentieth century and its theatre. 'Pervasive though [that negative vision] is', she writes, 'the specifically ecological meaning – as opposed to the mere theatrical presence – of this imagery has remained occluded, unremarked' within academic responses to such plays (p. 24). As the field of ecocriticism evolved, however, more scholars would come forward to 'remark upon' the potential ecological meanings of Beckett's writing. In 2010, Downing Cless argued (similarly to Chaudhuri) that 'In Beckett and beyond, nature is eviscerated' and that 'Most postmodernist theatre is post-nature' – but in the same book he also argued that 'a very positive goal of contemporary eco-theatre is to help audiences find a genuine garden and actual animals in a world of hyper-simulation', suggesting that directors could reimagine canonical plays (including those of Beckett) in ways that would make the ecological features of his *oeuvre* more apparent (p. 6). Also in 2012, Greg Garrard proposed that *Endgame* should not be seen as *occluding* the ecological but instead that it must be 're-read' as a 'precursor to ecocriticism' and therefore that it is 'the perfect play for the era of anxiety about climate change' (p. 383). Carl Lavery and Clare Finburgh would go even further in their 2015 edited collection, *Re-Thinking the Theatre of the Absurd*, by suggesting that 'Beckett demands a confrontation with … the "ecological repressed", the refusal of human beings to accept that their actions have environmental effects and, consequently, that their fate is bound up with that of the Earth' (p. 10). And, by 2021, Anna McMullan would write that 'Beckett's textual and audiovisual imaginings of ruined worlds on the edge of extinction are remarkably prescient for a twenty-first century confronting a global climate emergency' (p. 1).

That evolution of vocabulary over almost thirty years – from 'occlusion' to 'prescience' – reveals a transformation not so much in the production of Beckett's plays as in critical approaches to those plays. Scholars began with the ecological being seen as something that was hidden and ignored (as Chaudhuri proposed), before seeing it as a precursor to a more conscious form of awareness of the natural world (as Gerrard suggests), to viewing it as having always been present in Beckett (as McMullan implies).

Building on those approaches, in this first section I want to explore Beckett's drama in relation to the concept of ecodramaturgy, an idea developed by Wendy Arons and Theresa J May (though now used widely in the field). 'Ecodramaturgy,' writes May, 'is theatre praxis that centers ecological relations by foregrounding as permeable and fluid the socially-constructed boundaries between nature and culture, human and nonhuman, individual and community' (2021, p. 4). She goes on to define it as having three broad components:

> It encompasses both making work (making theater) and critical work (history, dramaturgy and criticism) in three interwoven endeavors: (1) examining the often invisible environmental message of a play or production, making its ecological ideologies and implications visible; (2) using theatre as a methodology to approach contemporary environmental problems (writing, devising, and producing new plays that engage environmental issues and themes); and (3) examining how theatre as a material craft creates its own ecological footprint and works both to reduce waste and invent new approaches to material practice. (p. 4)

The production that I explore in this section combines these elements of May's definition – but I also draw on Lisa Woynarski's extension of May and Aron's ideas into what she terms 'intersectional ecologies', which she sees as offering 'a way of interpreting ecodramaturgical practices [by] foregrounding marginalised perspectives and acknowledging the multiple social and political forces that shape climate change and related ecological crises' (2020, p. 6). These perspectives are used to consider the Irish-language version of Beckett's *Happy Days*, which was directed by Sarah Jane Scaife for her own Company SJ, and which was performed in 2021 on one of the three Aran Islands that lie off the west coast of Galway: a space at the edge not only of Ireland but of Europe – and a space, also, that occupies a central place in the Irish cultural imagination.

By being situated in a location that is geographically peripheral – while also being performed in an endangered minority language – Scaife's *Happy Days* (*Laethanta Sona* in Irish) creates a continuum between ecological crisis and longer-term social and political forces.[2]

The iconic status of the Aran Islands has its roots in the late nineteenth century, when they became an important location for folklorists, writers, and scholars dedicated to the revival of the Irish language (which, already in decline in the first half of the nineteenth century, had collapsed after the Great Famine, thus creating a link between ecological disaster and language usage that persists into the present). The architects of the Irish Dramatic Revival, Lady Gregory and William Butler Yeats, both visited the islands, but the plays of John Millington Synge are the best known examples of theatre that was either set there (as in *Riders to the Sea*, 1904) or inspired by stories and customs that had been gathered from the Irish-speaking communities there (such as *In the Shadow of the Glen*, 1903 and *The Playboy of the Western World*, 1907). The islands' status was also determined – probably over-determined, indeed – by Yeats, who placed them in a symbolically binary relationship with Paris, as the site of an authentic pre-modern culture that could offer Irish writers a firmer imaginative grounding than did the cosmopolitan centre of European naturalism and modernism. 'Give up Paris,' he urged Synge when they met for the first time in 1896. 'Go to the Aran Islands . . . Live there as if you were one of the people themselves; express a life that has never found expression' (Synge, 1982, p. 68). Those words came to be seen as a kind of unofficial summation of the ethos of the Revival – and they partly explain the later trajectory of Beckett who, in giving up Ireland for Paris, and in giving up the Irish Revival for European modernism, could not more thoroughly have rejected Yeats's advice to Synge.

Of course, as with so many of Yeats's other characterizations, the Aran Islands/Paris dichotomy was a deliberate over-simplification. Synge may have gone to Paris under the influence of French writers such as Loti, Baudelaire, and Anatole France but he also deepened his knowledge of Irish culture in that city by studying Celtic Civilisation with Henri d'Arbois de Jubainville at the Sorbonne – and it was that experience, as much as Yeats's directive, that inspired Synge to visit the islands for the first time in the summer of 1898. He returned each year until 1902, recording his impressions of the people, their lifestyle, and their language – and published many of those observations in his memoir and travelogue *The Aran Islands* (1907), gathering stories and sayings that remain as valued a source for folklorists as they are for scholars of theatre

---

[2] *Laethanta Sona* was a co-production between Company SJ and the Abbey Theatre, performed as part of Galway International Arts Festival 2021. My discussion is based on my attendance on 4 September 2021.

and literature. Occupied by an Irish-speaking peasantry whose livelihood was dependent upon fishing in the Atlantic, the three islands were impoverished and marginalized, their inhabitants subject to the vagaries of weather, tides, ocean currents, and other forces of nature. Synge admired many of the islanders' customs, was inspired by their resilience, enjoyed their stories, and befriended several of them – but he also showed an unsentimental determination to record the precarity and hardship of their lives.

The iconic status of the islands has been many times reiterated since Synge's death in 1909, in films, plays, novels, and other artworks. That centrality was reinforced when Druid Theatre staged a celebrated performance of Synge's *Playboy of the Western World* on Inishmaan in 1982, a visit that represented a kind of homecoming, bringing Synge's drama back to the location that had inspired the play (although set in the mainland Irish county of Mayo, *The Playboy* was based on a story that Synge heard on Aran). That visit demonstrated how the islands occupy a position of aesthetic centrality within the Irish theatre, such that performing a play there might be compared to what staging Shakespeare at Elsinore might be for the English theatre – which is to say that the location has importance not so much geographically, historically, or even spatially, but for the place it occupies in the culture's imagination.

The status of the islands within the culture needs to be contrasted with their socio-economic situation, however. Since the middle of the twentieth century, the islanders' livelihood has become more dependent upon tourism, much of it related to the growth of international air travel. And although the islands are now well connected to the mainland by ferry and light aircraft, their prosperity remains somewhat precarious. The population of the three islands stands at slightly more than 1,200 people as of 2022 (less than half the number that had resided there when Synge was visiting), and there are few signs of that figure increasing. They face other threats too. As predominantly Irish-speaking locations, the islands are part of Ireland's *Gaeltachtaí* – regions, mostly on the western seaboard from Donegal to Kerry, that have been excessively affected by emigration and impoverishment and which, because of their coastal locations, are also at a high risk from the sea-level rises that are already occurring due to climate change (Walsh, 2022, p. 221). This problem encapsulates a challenge faced disproportionally by indigenous communities globally: many are dependent upon tourism for their financial survival, but the carbon emissions associated with tourism represent a threat to their sustainability too (see Kronik and Verner, 2010).

There is no record of Beckett ever having visited the Aran Islands, though he did travel with his brother to the nearby mainland counties of Galway and Mayo in the early 1930s – perhaps inspiring the line in Lucky's speech about a 'skull in

Connemara' in *Waiting for Godot*. His *More Pricks than Kicks* (1936) describes that western region as a 'dream of stone', perhaps inspired by the stone walls that fill its landscape (2010, p. 135) – and a phrase that, as I will mention again below, was reanimated in the Irish-language version of *Happy Days*. But Beckett was certainly aware of the islands, not least through his appreciation for the plays of Synge. As James Knowlson writes, the 'revivals [in the 1920s] of ... Synge's plays at the Abbey were of greater significance to Beckett than the work of any other Irish dramatist. When I asked him who he himself felt had influenced his own theatre most of all, he suggested only the name of Synge' (1997, loc 1376). For this reason, the staging of Beckett on Aran reconnects him to a major precursor and influence.

The production that is the subject of this section was staged outdoors on Inis Oírr, the smallest of the three Aran islands, from 30 August to 5 September 2021, where it was called *Beckett sa Chreig: Laethanta Sona* ('Beckett in the Rock: Happy Days'). Although that production might have seemed like an imitation of the approach of Druid Theatre – which had not only staged *Playboy* on Aran in 1982, but had also presented *Waiting for Godot* in a field on Inishmaan in 2016 – it is important to observe that its director Sarah Jane Scaife has a long track-record of presenting outdoor performances of Beckett, notably in her *Beckett in the City* series that ran from 2009 to 2015 (see Scaife, 2016). Her production also went beyond Druid's in rooting Beckett not only in the unique landscape of the Aran Islands but also by translating the play into Irish.

The site for the production was a field called 'Creig an Staic' – translated as 'Beckett in the Rock' in the English-language version of the title, although (as often happens with Irish placenames) the original name conveys greater nuance and more historical resonance. Writing in the show programme, Cathleen Ní Chongaile observes that the placename came from a 'man who spent his time erecting a limestone wall around a "creig", crag in English, an area of rough ground consisting of flagstones and some fertile fissures' (2021, p. 2). This field is, she writes, 'located on the south western tip of the island where, every so often, gigantic Atlantic swells scale over the wide expanse of sea-made boulder barrier beside it to reclaim the creig' (p. 2). The Irish placename thus captures an arrestingly dialogic relationship between one human individual and the forces of nature. 'Stack's Crag' is a field encircled by stone walls, their purpose originally to mark his ownership of the land while also protecting the few 'fertile fissures' (gaps between the flagstones, between which grass and flowers grow) from ocean waves. 'Beckett in the Rock' doesn't capture any of these meanings. The Irish title, however, locates Scaife's production within an Irish eco-literary tradition, present within Irish poetry especially, called *dinnsean-chas*, which, as Seamus Heaney remarks, involves 'poems and tales which

Figure 1 Building Winnie's mound (photo by Cormac Coyne)

relate to the original meanings of place names and constitute a form of mytho-
logical etymology' (1980, p. 131). Beckett might have 'given up' Ireland for
Paris – but this setting showed how he could be brought back to an Irish context,
with new associations and continuities becoming visible through that process of
return.

Another continuity was established by the production: that between Beckett's
artistry and the craftwork of the local artisans who built a mound for Winnie
(Figure 1) from the limestone rocks that surround the field. Designed by the
sculptor Ger Clancy, the mound was built by Tomás Noel Searraigh, Máirtín
Celine Seoighe, and Máirtín Stiofáin Seoighe – all residents of the islands, who
were using a form of expertise that had been practised in Aran over several
generations, and which is also evident in the creation of the walls of the field
themselves. Synge took stories from the Aran Islands and used them to create
plays; here the islanders took a play by Beckett, and used it to create a new
object that fast became a source of local pride.

Other features of the design were also inspired by the landscape. The fabric
for Winnie's costume was chosen by Sinéad Cuthbert to develop a visual
association with the flowers that grow between the limestone flagstones,
which means that audiences would have seen a correspondence between the
non-human beings growing by their feet and the primary character in the play.
Winnie thus seemed almost like an outgrowth from the rocks, concretizing
decades of scholarship that had tracked Beckett's 'roots' from Europe back to

Ireland (see for example Kiberd's comparison of Beckett with the Irish-language writer Máirtín Ó Cadhain in *Irish Classics*, 2001).

This setting also had an impact upon the audiences' reception of the work. The islanders were of course encouraged to attend the show, but the number of tickets that were made available for sale far surpassed the population of Inis Oírr, thus showing how it was conceived of as an experience to which audiences would, in general, need to travel to. The journey to Inis Oírr by boat takes slightly more than an hour, and brings travellers from Galway Bay out to the open Atlantic, where mild seasickness is not uncommon. Upon arrival at a small harbour, visitors then had to walk uphill towards Áras Éanna, the island's arts centre and the first assembly point for the audience. While there, the audiences were encouraged to view an exhibition in the centre's gallery: entitled 'Laethanta Sona – Saol ar an gCreig', the exhibition celebrated the women of Inis Oírr through recordings and photographs, thus preparing the audience to see a correspondence between Winnie in the play and the women of the islands (about whom Synge wrote extensively in his own *The Aran Islands*).

From Áras Éanna, the audience was guided along a walk of roughly 2 kilometres to Creig an Staic, mostly through rough paths that were bordered by the island's characteristic stone walls. As can be seen from the production photograph in Figure 2, the performance space is itself surrounded by those stone walls, and the ground is covered by limestone flagstones that have grass and flowers growing between them. This setting is different from any that Beckett had envisaged for the play, yet seems faithful to his call for a performance environment that displays a 'Maximum of simplicity and symmetry' (1961, p. 7).

Audience-members sat on rocks that had been carefully positioned around the space; cushions were provided, but audience comfort was not the primary priority. That outdoor setting had several other impacts upon the audience's capacity to receive the action. Since the location was free of trees and peaks, it was also relatively lacking in shelter from the elements: anyone watching the play would therefore have needed to contend with wind, rain, sunlight – or (being in Ireland) with all three simultaneously. The lack of shelter also meant that there were fewer surfaces than would ordinarily be found in a theatre building – so, rather than projecting her voice, Bríd Ní Neachtain (who played Winnie) instead needed to wear a mic, her voice amplified by speakers placed, somewhat inharmoniously with the surroundings, at the sides of the field. The actor's voice thus required technology to overcome the conditions of the performance space, an example of the human being dwarfed by the natural setting that is not dissimilar to what happens when those stone walls fail to withstand the swell of the nearby Atlantic.

Figure 2 *Laethanta sona* (photo by Cormac Coyne)

Happy Days has been staged and received in ecological contexts before. In a review of Trevor Nunn's production of the play in 2021, for example, Jonathan McAllister detected evidence both in the staging and in Lisa Dwan's performance of the 'cognitive dissonance' that is a characteristic of watching Beckett 'in an age of climate catastrophe' (2022, p. 226). Catherine Love has also written about how Katie Mitchell's 2015 production of the play for the Deutsches Schauspielhaus in Hamburg 'attempted to convey the stark realities of climate change by bringing the destructive forces of environmental catastrophe into the space of the theatre' by placing Winnie up to her neck in water rather than sand (2020, p. 234). There has also been a novel, Claire Thomas's *The Performance*, which uses a staging of the play to address themes of gender, community, and environmental destruction in Australia (2020). The Irish actor Fiona Shaw has also written about how her performance as Winnie acquired tragic force when its 2008 visit to Greece coincided with major forest fires in that country, giving rise to a disturbing 'alignment between the burning images in *Happy Days* and the fires raging through Greece'. Shaw explains how she –

> thought of Winnie's lines: 'Is it not possible, with the sun beating down, and so much fiercer down, things to go on fire never known to do so in this way, I mean spontaneous like.' That night, the ancient theater [in Epidaurus] had the quality of an etching of the human brain, a tiered slice that fills the eye. I felt the slide rule of time did not apply. I was entirely in the past and in the present at the same time – all the people who had sat there, in togas two

millennia ago, now in T-shirts and jeans – but I felt the frisson as I hit the lines: 'Might I myself not melt in the end, or burn, I do not mean burst into flames, but little by little be charred to a black cinder, all this visible flesh'. The following morning, in the hotel, local tourists were weeping. The hills outside Athens were burning, and within an hour our second performance was cancelled, as were all performances in Greece. Official mourning was declared. Theatrical tragedy had given way to the real thing. (2008, p. 112)

Shaw's encounter with environmental collapse disturbed the distance between the contemporary and ancient periods of Western theatre, locating her 'in the past and in the present at the same time' as she acted Beckett's play. Here again it seems possible to identify evidence of how theatrical form struggles when faced with the task of apprehending and representing the vast timespans that are encompassed by such hyperobjects as geology or weather.

Scaife's production was not performed against the backdrop of an environmental tragedy in the way that Shaw's was, but it did employ methods that, rather like the experience that Shaw encountered in Greece, broke down audiences' sense of time and space. From a reading of Beckett's script, one might form the impression that *Happy Days* is a play that intends to put audiences *in medias res*: they are likely to assume that the day before the play begins was much like what they see (or read about) in Act One, and that the day following the completion of the play will be much the same as what they have encountered in Act Two, with Winnie slowly disappearing beneath that mound of sand. But for this production, the boundaries around the staging of the play became more uncertain – since the experience necessarily involved a long walk (or, for those with mobility difficulties, a short drive) from the assembly point and, for most audiences, a journey by boat. The audience's reception of the production was thus linked to climate, environment, and their own embodiment: it was not so much the production as the audience themselves that were site-responsive. Many of the performances were accompanied by the sound of Atlantic waves, and many by birdsong; some viewers may have watched the play with the lingering effects of seasickness; some might occasionally have felt uncomfortably warm as the glaring sun reflected against the bright limestone. In such contexts, it seems more difficult to claim that the experience begins only when the ringing of a bell summons Winnie into speech.

Critics often comment upon the repetitious quality of Beckett's dramaturgy, but on Inis Oírr, repetition seemed more like something cyclical than the 'dwindling' that is one of the central features of his writing for the stage. By being performed on land partly given over to farming, this production was in tune with the human intervention into the passing of the seasons through agriculture; by requiring a journey by boat, it was aligned with the tides.

'Everyday life,' writes Alan Read, 'is bound not only by the cyclical return of natural phenomena, the seasons, elements and climate ... but more mundanely by the linear rhythms of life's trajectories through birth, marriage and death.' He goes on to argue that 'any analysis that privileges one cycle over the other is destined to miss the subject of those patterns' (1995, p. 117). *Laethanta Sona* was attuning audiences to the place of natural cycles within their own lives, placing the linear and the cyclical into correspondence with each other. There was also an opportunity for audiences to think about the cyclical nature of theatrical influence: Synge heard stories on the Aran Islands that found their way into his play, and those plays influenced Beckett. Hence, bringing *Happy Days* to Inis Oírr can be seen as an example of an influence influencing itself, a theatricalized version of the ouroboroses that Timothy Morton writes about in *Dark Ecology* as an iconic expression of ecological awareness (2016, p. 59).

May's discussion of ecodramaturgy provides tools with which to understand this production. Scaife's production shows, first, how it is possible to detect 'invisible ecological ideologies' in Beckett's stagecraft and dialogue (2021, p. 4) – the awareness of which is heightened by the fact that, like Winnie herself, the audience is trapped in a place where they are exposed to the weather: if audience-members feel uncomfortable as they shift around on the stone seats and try to avoid being sunburned, then they are starting to understand Winnie's situation physically in a way that is less likely to arise in a theatre building. It is also a production that uses theatrical form – including the meta-theatrical processes around the staging of the script (the walk, the boat-trip, the use of rocks as seats for the audience) to force an encounter with environmental phenomena such as weather and coastal erosion as well as more benign events such as the unexpected interventions from birds and other animals that sometimes appeared in the performance space. *Laethanta Sona* also provides an example of how theatre 'as a material craft creates its own ecological footprint' – because it used stone (and knowledge) from the island in the construction of the set. That action literalizes the metaphor of bringing something buried to the surface, but it also demonstrates the efficacy of practice that 'works both to reduce waste and invent new approaches to material practice'. Not irrelevantly, it also generated paid work for members of the island community, which is more than can be said for the Aran islanders whose stories had inspired Synge's dramas a century previously.

All these approaches are relevant for the present argument – but of greatest significance for this argument is the staging of the play in the Irish language, thus bringing Beckett into dialogue with the indigenous culture of Ireland. As Lisa Woynarski argues, 'The inherent interconnections between colonialism and ecological subjugation are too prevalent not to be considered by anyone working in ecology and art, and as Maori scholar Linda Tuhiwai Smith writes, the responsibility to make change is "in both the non-indigenous and indigenous worlds'" (2020, p. 184). Woynarski also refers to the importance of Sandy Grande's *Red Pedagogy* (2004), which 'seeks out new epistemologies and paradigms based on Native American knowledges in education, which can open up and diversify critical theory in the academy, not towards assimilation of knowledge but towards decolonisation, recognition of difference and multiple ways of knowing' (quoted by Woynarksi, p. 183). Scaife's production can be seen as coming from a similar (though certainly not identical) desire to use the indigenous language and culture of Ireland to stimulate new forms of knowing, new ways of recognizing difference.

Beckett did not speak Irish but translations of his plays into that language were proposed from a relatively early stage of his playwriting career. Those translations had an emphatically postcolonial ethos from the outset. Shortly after the Irish premiere of *Waiting for Godot*, for example, the actor Cyril Cusack attempted to produce a bilingual version of the play that would feature Vladimir and Estragon delivering their lines in Irish while Pozzo and Lucky spoke English. That idea that never made it to the stage but it can be seen in adapted form in the Walter Asmus productions of *Godot* that dominated the Irish stage from the 1980s onwards: in those productions, Didi and Gogo speak in Hiberno-English while Pozzo and Lucky speak in Anglo-Irish accents or English received pronunciation, thus contextualizing the play's exploration of power dynamics in relation to both social class and the history of Anglo-Irish conflict. As such examples demonstrate, the status of English in relation to Irish thus has colonial, postcolonial, and political consequences that are likely to inform decisions about translation while also affecting the reception of Irish, Anglo-Irish, and/or Hiberno-English productions of Beckett's plays. These contexts may not be legible (or even interesting) to audiences outside Ireland, but I mention them to highlight how the ecodramaturgical approaches of May can be extended to include other historical and cultural matters.

One way of illustrating the relationship between the Irish and English languages – and thus to understand the significance of this production – is to mention the fact that the word for the English language in modern Irish is *Béarla*, a word that originally could refer to any language, but which gradually came to mean only the speech of Ireland's colonial ruler. As Arbuthnot, Ni Mhaonaigh, and Toner explain (2019), *Béarla* originally came in the form '*bélrae*' (a word that shares roots with the Irish word for a mouth, *béal*)' (p. 183). The transformation of the word from meaning 'language in general' to something like '*that* language' allows one to trace the impact of English hegemony upon Ireland through semantic change. It may be interesting to observe in this context that most modern Irish nouns that denote languages are gendered as feminine, including *Gaeilge*, the word for the Irish-language itself, as well as the word *teanga*, which represents the word 'tongue', both in the sense of the bodily organ and as a synonym for 'language'. *Béarla* is gendered masculine, however – again providing evidence of how political power affects and/or reflects the construction of language. As I discuss in the next section, the gendering of Anglo-Irish relations as a patriarchal conflict between a masculine England and a feminine Ireland has had theatrical as well as linguistic manifestations.

Happy Days is in its own way an exploration of the construction of gender, as has long been understood thanks to the scholarship of such writers as Linda Ben-Zvi (1992) and Elin Diamond (2004), among many others. Its translation into Irish thus forms a link between some of the play's themes and the language that is being used to express those themes – something that is not possible in its English-language version. Its French version *Oh les beaux jours* is written in a gendered language too (a feature I will mention again briefly below) but not necessarily in a way that maps onto political or historical relations in the way that Gaeilge does.

The play's translation is by Micheál Ó Conghaile, a publisher and playwright. In many respects, his version is very faithful to Beckett's English. For example, the French text calls in its opening stage directions for 'une toile de fond en trompe-l'oeil très pompier' (1963, p. 11) which in English is '*Very pompier trompe-l'oeil backcloth*' (1961, p. 7). In Irish, Ó Conghaile renders this as 'Culbhrat í stíl trompe l'oeil le cur i gcéil' – that is, he retains the French 'trompe l'oeil', just as Beckett did, but translates the English words (2021, p. 15). Such fidelity has the impact, however, of drawing notice to a small number of deviations, perhaps the most prominent of which is a decision to retain the English language whenever Winnie reads aloud any descriptions of products or when Willie reads advertisements from his newspapers. When Winnie reads her toothbrush's product description in the English version of the play, the line in the script is '*guaranteed ... genuine ... pure ... what?*' (1961, p. 10). In French, Beckett uses words that have close English analogues but are also used in French and which would therefore probably not be seen as inherently foreign to Francophone audiences: 'véritable ... pure ... quoi?' (1963, p. 14). However, in Ó Conghaile's version, the only word to be translated into Irish is the question that corresponds to 'what' or 'quoi'. 'Genuine .... pure ... céard?' (2021, p. 17). This exemplifies the playful nature of this translation: in refusing to render 'genuine' and 'pure' into Irish, Ó Conghaile creates a version of the play that might seem *less* genuine, *less* pure, than the original. He could have translated those words into such Irish equivalents as *fíor* or *íon*, but instead chose to signify how English retains a presence in the play in the language of advertising and the mass media.

This strategy calls to mind a line in Brian Friel's play *Translations* (1980), which is also about the relationship between Irish and English: 'I explained that a few of us did, on occasion [speak English],' one of Friel's characters remarks, relating a conversation he'd had with a member of the British army. But English, the character says, is 'usually for the purposes of commerce, a use to which his tongue seemed particularly suited' (2013, p. 269). The status of English as invasive in relation to Irish thus becomes more evident in the translation (or, strictly speaking, the non-translation) of commercial language in *Laethanta Sona*. 'As soon as we are told that a product is "natural",' notes Bruno Latour, 'we understand clearly, at worst, that someone is trying to trick us and, at best, that someone has discovered another way of being artificial' (2017, p. 21). Ó Conghaile's script gives verbal form to that observation.

Actors performing this role in English or French might interpret the 'what' or 'quoi' at the end of those lines in a variety of ways: perhaps as a sign of Winnie's fading eyesight, perhaps as an attempt to ironize the products' claim to be 'pure', perhaps in some other way. Ó Conghaile's use of the English original – followed by the Irish *céard* – generates the possibility of a new approach to the acting: that of the native Irish speaker who is encountering an English that she does not fully understand. In such a context, Winnie's later declaration of an intent to 'speak in the old style' (In Irish '*le lahbairt ar an sean-stíl*', p. 24) seems to have a new political force, calling to the audience's notice the long history of suppression of the Irish language, which could itself be termed 'speech in the old style'. None of this can or should be explained in relation to Beckett's authorial intention, but nor is it imposing a meaning upon the play that is entirely absent from it; again, Bould's notion of an Anthropocene Unconscious seems helpful in this context.

There are also meaningful choices in the application of grammatical conventions. As explored in the final section in the discussion of the origins of the word *turlough* – a compound of the Irish words for 'dry' and 'lake' – the Irish language includes a feature of bringing distinctive (and sometimes apparently contradictory) words into alignment with each other in a single noun, a device employed by Ó Conghaile when he renders 'old things' and 'old eyes' from the English script (1961, p. 12) as the single words 'seanrudaí' and 'seansúile' respectively (2021, p. 19). The French original presents the same lines as 'Vielles choses' and 'Vieux yeux' (1963, p. 17). In Irish, the word 'rud' (or 'thing') is masculine and súil ('eye') is feminine; in French it's the other way around, but in both cases the blend of gendered nouns can be aligned with the treatment of gender in the play itself. Ó Conghaile could have presented these words as separate from each other ('sean rudaí' would in some ways have been a more standardized translation from the English), but the employment of the device reminds one of the Irish language's capacity to capture the way in which objects can be so modified by their contexts as to become almost like new beings: Winnie's 'oldthings' and 'oldeyes' are presented as categorically and ontologically different from their separate selves.

The playful negotiation of Irish and English world-views is evident in other ways. Early in the original script, Beckett refers to a town in England called 'Borough Green', which Ó Conghaile translates as 'Glas na Boróige' (literally, 'the green of the borough', p. 22). The last word in the phrase is a Gaelicization of the word 'borough'. The Irish word *buirg* could have been to render the placename literally, but by reproducing the English sound in the Irish language, Ó Conghaile is doing to an English placename what British colonialism had done to Irish placenames. As many theatre scholars will be aware from their knowledge of Friel's *Translations*, English rule in Ireland involved a process of re-mapping the country in the early nineteenth century, turning Irish place-names – with their long history of communal and environmental associations – into meaningless English sounds. That act of renaming had the impact of placing a barrier between the inhabitants of a place and their local and environ-mental histories. For example, residents of Dublin, Ireland's capital and Beckett's birthplace, are sometimes unaware that their city takes its name from two Irish words, for 'dark' and 'pool', which originally referred to a tidal pool that was located on what are now the grounds of Dublin Castle, the place from which English colonial rule was exercised over the island. The Anglicized 'Dublin' is comparatively and (it could be argued) deliberately meaningless. Ó Conghaile's 'Glas na Boróige' has a similar impact upon Borough Green, placing a barrier between the Irish audience-member and the actual English town.

The Irish translations, along with the sculpted stone mound, allow for a consideration of the ways in which Scaife's production aligns with indigenous performance practices internationally. Helen Gilbert has written about how 'Communal memory, a key concern in many indigenous societies, builds contingently from ... knowledge systems, reiterating the embodied basis of cultural transmission' (2013b, p. 174) – and this is a production that demon-strates how communal memory can be traced through the performance of Beckett by means of language, placenames, literary influence, and the passing down through successive generations of knowledge of artistic craft. Denise Varney writes about how some of the most compelling Australian work about climate change is 'undertaken in the Australian performing arts by Indigenous theatre companies representing a cultural perspective that rejects the Western mastery of man over the environment, of the separation of human and non-human, and that recognizes the unique life force of land or country' (2022, p. 7). Beckett's writing is *of* the West but it too rejected the Western notion that 'man' was master over nature. But Scaife's production also does much to bring that feature of his work into alignment with the 'unique life force' of the Inis Oírr setting.

My aim in this section has been to explore how the ecodramaturgical approach developed by May and others can be applied not only to the analysis of a performance, but also to the making of performances. By rendering Beckett's 'dream of stone' in material form, Scaife situated *Happy Days* in the Aran landscape – but also repositioned him, and the islands, in the cultural imagination of the nation. That approach illustrates how the Irish colonial context allowed for new correspondences to emerge between ecological crisis and earlier phases of human history. That topic becomes the subject of the next section, which attempts to show how theatre can be used to explore the construction of the Anthropocene in the early modern period.

2 Deepening Time: Shakespeare's *A Midsummer Night's Dream*

When Shakespeare's actors first delivered his speech about how 'all the world's a stage', they were standing in a theatre that might have been built with Irish timber.

Those words are not from *A Midsummer Night's Dream* (a production of which is the primary subject of this section) but from *As You Like It,* a play believed to have been premiered in 1599, perhaps as one of the first dramas to appear at the then newly opened Globe Theatre. That 'wooden O' was constructed from timber that had been used in an earlier building called 'the Theatre', which (according to legend) Shakespeare's company had disassembled during the winter of 1598, carrying it across the frozen Thames to the city's south bank, where they had built their new venue (see Shapiro, pp. 1–9). That act of repurposing – one could even call it recycling – needs to be seen as arising from the Burbages' strained finances but, as Vin Nardizzi has pointed out, it must also be understood in the broader context of 'economic and ecological factors pertaining to the health of England's woodlands' (2011, p. 55). Timber was scarce in Elizabethan England, a fact that made the construction of a theatre from second-hand materials a necessity (for a full discussion, see Steffen, 2023).

That scarcity of timber also explains the growth of English interest in Irish trees during the Tudor period. As Thomas Herron (1998) shows, imperial covetousness is traceable in the writings of Edmund Spenser, whose poetry 'shows a fondness for idyllic Irish forests' (p. 5). Spenser, however, 'also saw the wisdom of cutting [Irish forests] down, so as to creatively channel destruction and build a more civilized political landscape' (p. 5). For Spenser, as for many English people living in sixteenth-century Ireland, the forests were a place that 'harbored "Wolves and Thieves", hostile wood-kernes and other Irish rebels'; their destruction was thus justifiable not only for profit but also for 'security reasons' (p. 5). Similar imagery related to the Irish finds its way into many of Shakespeare's plays – from Richard II's angry description of the Irish as 'rough rug-headed kerns' (2.1.156) to Rosalind's dejected dismissal of Orlando's verses as 'like the howling of Irish wolves against the moon' in *As You Like It* (5.3.105).

Irish timber had been exported to England since at least the fourteenth century, its expropriation (and sometimes its destruction) justified in relation to 'security reasons' – as for instance when the real Richard II, campaigning in Ireland in 1399, 'ordered the mobilization of 2,500 "country people" to cut and burn some of [an Irish rebel's] woods' (Everett, 2014, p. 21). In the two centuries that separate Richard's act of despoilation from the opening of the Globe, Irish timber became commonplace in English construction, often used for its navy, not to mention its churches and houses. Why not its theatres too?

Such concerns might seem obsolete now, but the destruction of Ireland's forests and woodlands remains a potent subject in Ireland. In the cultural history of that nation, the loss of forests and woodlands has been ascribed to English misrule – and it is a loss that continues to be felt a century after independence, when Ireland has the second-lowest rate of forest coverage in the European Union. In his study of Irish woodlands, Nigel Everett (2014) highlights the need for historians to adopt a more nuanced approach to the narrative that a 'spiteful' Queen Elizabeth I ordered an 'arboreal holocaust' in Ireland, both to deny rebels a hiding place and to exploit the value of that natural resource (p. 11). Nevertheless, that narrative is widely believed, resonating into the present in many ways. Generations of Irish schoolchildren have studied the poem 'Caoineadh Cill Cáis' ('the Lament for Kilkash') in their Irish-language lessons, internalizing the idea that the destruction of Ireland's woodlands and its loss of sovereignty to England should be seen as metaphors for each other. And as recently as 2020, the Irish film *Wolfwalkers* (nominated for best animated feature at the 2021 Academy Awards) generated much of its drama from the correlation of English rule with the destruction of Irish forests and the hunting to extinction of Ireland's wolves during the seventeenth century.

Links between ecology and colonialism have been commonplace in Irish culture for centuries, especially in Irish-language poetry. Lucy Collins has explained how the changing weather patterns that occurred at the end of the Little Ice Age inspired Irish poets to explore political change, something that continued up to and beyond the Romantic period. Those poets, she writes, engaged with the weather not just metaphorically but experientially, using climactic instability as a way of thinking about their loss of sovereignty (2021, p. 347). Such ideas find their way into modern Irish writing also: Joyce's *Ulysses* (1922) has Bloom's antagonist the Citizen link the fate of Ireland's forests with its degraded national status: ' – Save them, says the citizen, the giant ash of Galway and the chieftain elm of Kildare with a fortyfoot bole and an acre of foliage. Save the trees of Ireland for the future men of Ireland on the fair hills of Eire, O' (2000, p. 423). Joyce might be mocking the Citizen's nationalism in that passage, but he is also showing the potency and persistence of the link between nature and nation.

Nardizzi argues that, when Shakespeare's actors first performed his plays, they would have gestured towards the wooden structure of the playhouse to realize in the audience's imagination the natural world being evoked by the language:

> Shakespeare routinely conscripted the woodenness of the playhouse to per-form the role of tree, woods, forest, orchard, and park. When characters invoke such settings, their words usually indicate that they also physically motion toward some thing. This thing could be a tree stage prop, but it could just as easily have been one of the wooden (and perhaps painted) posts supporting the stage canopy. (2011, p. 51)

Gabriel Egan has contested that claim by pointing out that 'open-air amphitheatres were often constructed so as to disguise their wooden origins', one example of which, he shows, is the desire to make the stage pillars 'shine like marble' (2016, p. 31). Nevertheless, he agrees with Nardizzi that there are 'stimulating analogies' between the fictional location and the 'real site of performance' (p. 31). How might those analogies have affected the reception of the plays?

'All the world's a stage': implicit in that phrase is the idea of the stage as a blank space, a location shorn of history, context, or roots – and implicit also is an apparent assumption that the geographical or colonial origins of the timber used to construct the stage are unworthy of attention. Similar links between the materiality of the stage and the colonization of Ireland are present in another Shakespeare play believed to have been first performed in 1599 – *Henry V*, which famously begins with its Chorus's description of what one might now call the phenomenology of theatre spectatorship. The stage is an 'unworthy scaffold' that brings forth 'so great an object' as 'the vasty fields of France', says the Chorus (Prologue, 11–12). The Globe's newly built wooden walls confine 'two mighty monarchies,/ Whose high upreared and abutting fronts/ The perilous narrow ocean parts asunder' (20–22). This is a space in which the actors' speeches about horses can allow audiences to visualize 'their proud hoofs i' the receiving earth' – a space that can change their perception of the flow of time, 'Turning the accomplishment of many years/Into an hour-glass' (30–31). Thus, the stage space is not made of materials taken *from* the world depicted on stage; it is more like a substitute *for* that world.

Might there be an erasure at work here, a transformation of something into an apparent nothing? Even if those wooden pillars looked like marble, the audience should have been aware that they were once living trees, and those audiences would also have seen in the Globe a structure built with a scarce commodity. From *Merchant of Venice* to *Much Ado about Nothing* to *Hamlet*, *King Lear*, and beyond, Shakespeare displayed an understanding that what is called 'nothing' is often a 'something' in disguise but, even so, the description of the Globe as a 'wooden O' – as a zero, a nothing, a blank – risks the effacement of that structure's natural and (possibly) colonial origins. After all, even if the Globe was not built with Irish timber, the scarcity of that resource was often used to justify the colonization of Ireland, a task that was well underway when *As You Like It* and *Henry V* were first performed. In that context one might consider how, later in *Henry V*, the Chorus turns his attention from the audience's interpretation of theatre to attempting to shape their responses to war in Ireland, by linking Henry's victories in France first to the imperialistic might of Caesar and then to the anticipated return from Ireland of Essex, 'the general of our gracious empress ... from Ireland coming/ Bringing rebellion broached on his sword' (5.0. 30–32). Time collapses in that speech – not in the 'accomplishment of many years/Into an hour-glass' that is enabled by dramatic form (as imagined by the Chorus at the beginning of the play) but in the conflation of Ancient Rome with Elizabethan England, the conflation of Caesar with Essex, the conflation of Gaul in the first century BCE with Gaelic Ireland in 1599. These words might be seen as evidence of an awareness of a kind of 'deep time' that is different from the one that scientists write about: the Chorus's speech is not describing the long span of knowledge that is captured by biology or geology but the 'eternal now' that is one of the core elements of imperialist and totalitarian imaginaries.

The ideas of Kathryn Yusoff are again pertinent here. Her *Billion Black Anthropocenes* explores how the rhetoric associated with geology can obscure histories of colonialism, racism, and white supremacism. 'Origins are not solely about geography', she affirms: 'They pertain to the question of how matter is understood and organized, as both extractable resource and energy, mobilized through dehumanizing modes of subject' (p. 66). The above speeches from Shakespeare's plays of 1599 show how matter, origins, and subjectivity were in the process of being reconfigured at precisely that time – and how Ireland and

Irishness played a role in that process. It is important to acknowledge that Yusoff's ideas are only partially applicable to Irish contexts: Irish people, after all, participated in British imperialism, settler colonialism, American slavery, and other forms of extraction and dispossession. But it is valid too to recall how Ireland was, as Declan Kiberd and others have argued, a 'laboratory' in which many of those practices were first perfected (1995, p. 1). The (possibly) Irish origins of the timber in Shakespeare's theatre-houses could thus be seen as one exemplification of Yusoff's consideration of how material things are manipulated through language to render oppression invisible. The Anthropocene, colonialism, and the early modern theatre all have 'origins' that can be traced to this period – and it is essential to disentangle their roots.

If one imagines plays such as *As You Like It* as being created at the outset of the Anthropocene era, then it might be possible to start tracing a relationship between theatre, the broader culture, and the construction of attitudes both to the natural world and to colonial expansion. My aim in this section, then, is to explore how theatrical revival can allow some of these correspondences to become more sharply visible. I discuss a production of *A Midsummer Night's Dream*, which was performed outdoors in the courtyard of Kilkenny Castle – which itself is a symbol first of the Norman and then of the English conquest of Ireland. By showing how weather, history, setting, and live performance interact in this production, I want to consider theatre's capacity to offer a deeper understanding of time and the environment for the Anthropocene.

Kilkenny is located in the south-east of Ireland, and is designated as a city not because of its size or mode of government, but because of its historical status: in 1609, it was granted a royal charter by King James, who was also, of course, the patron of Shakespeare's company at that time. It dates to a monastic settlement in the sixth century CE, but its growth increased with the invasion of Ireland by the Normans in 1169, after which time Kilkenny Castle was constructed – along with the city's walls, much of which still stand. It soon became a site for the Anglo-Norman dominance of Ireland outside of Dublin: in 1367, laws were passed that aimed to halt the decline of English rule in Ireland by outlawing Irish customs in favour of English ones – and those laws were called the Statutes of Kilkenny.

Kilkenny soon became one of the first sites of regular theatrical performance in Ireland. As Chris Morash recounts, its theatrical tradition extends

> back to at least 1366 when Archbishop Thomas forbade 'theatrical games and spectacles' on church property. On 20 August 1553, two plays were staged at the Market Cross in Kilkenny, *Gods Promises, A Tragedy or Enterlude* and *A briefe Comedy or Enterlude of Johan the Baptyses preachynge in the Wylderness,* by John Bale. (2002, p. 10)

Those plays were presented, Morash states, as an attempt to counter Kilkenny's 'vigorous Catholic mystery cycle tradition' (p. 10), showing how theatre in Ireland became a kind of proxy battlefield for the competing priorities, first of Irish and English, and then of Catholic and Protestant, communities. That overlap between theatre and nation became even more contentious in the 1640s, when the English Civil War caused Dublin-based Royalists to flee to Kilkenny. As a way of advancing their political aims, that group staged and published a new play in the city: *A Tragedy of Cola's Fury, OR, Lirenda's Misery* ('Lirenda' is an anagram of Ireland). That confederacy used its Kilkenny base to exercise authority over much of Ireland until the arrival of Oliver Cromwell's New Model Army in 1649. During the following four years, Cromwell's soldiers displayed a brutality that was unprecedented at that time in Anglo-Irish conflict, massacring large numbers of people and transporting thousands of others to the Americas as indentured servants. Again, Kilkenny occupies a pivotal role in that narrative.

The city re-enters Irish theatre history in the first two decades of the nineteenth century, when it became the home of a regular series of amateur performances of Shakespeare's plays. These performances were justified on the basis that they were not 'mere' entertainments but were charitable works by members of the Anglo-Irish gentry (surplus funds from ticket sales were donated to the poor). But the overall ambition of the productions is evident from the care and rigour with which the Kilkenny theatre space was constructed. As Michael Dobson writes (2011), 'the social and artistic assumptions of the project were built into the very shape of the Kilkenny theatre's expensively refitted and remodelled auditorium . . . [which] effectively excluded the lower orders altogether since nearly all the auditorium consisted of boxes', thus limiting the audience to the aristocratic actors' 'rich and polite social equals' (p. 55).

Those audiences may have been small in number but the productions them-selves were of sufficient impact that they may have been the subject of a rather mean-spirited satirical attack in London in 1809 when the English actor William Oxberry published a false playbill for a supposed 1793 performance of *Hamlet* in Kilkenny. The playbill describes the city as a 'learned *Matrapolish*', with the presentation of the word 'metropolis' in a phonetic version of the Irish accent an obvious attempt to mock that location's civic pretentions, while also repeating a common characterization of Irish speech as a distorted form of Standard English (Shakespeare himself had used this device in *Henry V* when he has Macmorris, his only Irish character, pronounce the word 'is' as 'ish' – 3.2.125). The role of Hamlet would be played by a Mr Kearns, the playbill announced; he would also perform several solos on his bagpipes. Ophelia would sing several 'favourite airs', including the suggestively titled 'We'll be Unhappy Together'. The roles of

Claudius and Gertrude would be cut altogether, upon the advice of a Revd. O'Callaghan who had deemed the characters 'too immoral for any stage'. But perhaps the most amusing statement was about the play's authorship: *Hamlet* was not written by Shakespeare but by a Limerick man called Dan Hayes.

These statements were obviously intended to be funny: Oxberry's readers would have regarded the suggestion that Shakespeare did not write his own plays as self-evidently silly. But that joke has sometimes fooled people: there have been at least three books that mention this performance as if it took place in the manner described; some of those publications' authors also appear to believe that the playbill's disputatious attitude to Shakespeare's authorship was taken at face value by Irish audiences. This example serves as a metaphor for the history of Shakespearean performance in Ireland: it was often provisional and subject to misunderstandings, and it occasionally became the punchline of an English person's dodgy joke. Kilkenny's role in that history arises because it so often was a place for the ruling elite to use theatre to dramatize a desire to achieve domination over the island, with the performances of Shakespeare central to those activities.

Those historical contexts explain the Kilkenny Arts Festival's decision to revive the tradition of Shakespearean performance at outdoor locations in the city, which began in 2012 with a visit by Shakespeare's Globe with *As You Like It* (with the Irish actor Deirdre Mullins playing Rosalind). The Globe returned in 2013 and 2014 with *The Taming of the Shrew* and *Much Ado About Nothing* respectively, with those productions easily transferring from the outdoor setting of the modern-day Globe to the courtyard of Kilkenny Castle. All three were well received but garnered relatively little critical analysis or attention within Ireland; instead, several newspaper features wondered rather fretfully why Shakespeare needed to be imported at all. Watching the Globe's *As You Like It* in 2012, Fintan O'Toole stated that he found the production enjoyable but noted how it reinforced his awareness of an 'embarrassment' in Ireland about that country's 'relationship with Shakespeare' just as there is sometimes an embarrassment in England about Shakespeare's attitude to Ireland. 'It is now a given that almost every culture has its own version of Shakespeare,' O'Toole complained. But 'Ireland doesn't' (p. 8).

That statement might explain why the Kilkenny Arts Festival began to co-produce Irish productions of the plays – a process it initiated in 2018 when *A Midsummer Night's Dream* was staged in the courtyard of Kilkenny Castle by Rough Magic. That theatre company was founded in Dublin in 1984, initially with the remit of producing the Irish premieres of contemporary international plays, though the influence of Shakespeare is also detectable in the use of a phrase from *The Tempest* as the company's name (5.1.50). Its artistic director

Lynne Parker quickly diversified her company's practice after its foundation, however, gaining an international reputation not only for her support of new Irish writers, but also for her successful staging of large-scale plays from the European repertoire, including Congreve's *Way of the World* in 1993, Sheridan's *School for Scandal* in 1998, and Schiller's *Don Carlos* in 2007. She also directed David Tennant in *The Comedy of Errors* at the Royal Shakespeare Company in 2000.

One of her most significant productions, however, was a 2006 Rough Magic *Taming of the Shrew*, which – although retaining the play's original setting and leaving the original script mostly uncut – used design and costume to suggest that the action was happening in a version of rural Ireland in the 1970s. That impression allowed the actors to deliver lines in rural Irish accents – producing sounds, rhythms and intonations that would have been more likely to appear in productions of classic Irish play by Synge or Lady Gregory. The company's use of the Irish voice to perform *Shrew* was considered revelatory: for most of the twentieth century, Irish productions of Shakespeare had been presented in a style that mimicked English models, making Irish productions seem both derivative and inauthentic. There was also a long tradition of Irish Shakespearean actors who had disguised their national origins by adopting an accent different from the one they had grown up speaking – including Peter O'Toole, the Irish actor whose Hamlet opened the British national theatre (see Croall, 2018, p. 13), and the Belfast-born Kenneth Branagh – whose status in the 1990s as Laurence Olivier's heir sometimes obscured his relationship with his place of birth. Irish accents appeared only gradually in Irish stagings of Shakespeare, first in a production of *The Comedy of Errors* at the Abbey Theatre in 1992 (see Lonergan, 2015), and then with Parker's 2006 production. She would return to Irish-accented Shakespeare with a production of *Macbeth* at Belfast's Lyric Theatre in 2011, and her *Midsummer Night's Dream* represented the first of a projected three-part cycle of outdoor Shakespearean performance that was to be performed at the Kilkenny Arts Festival from 2018 onwards (followed by *Much Ado* in 2019 and *The Tempest* in 2022). By that time, her reputation as a director of Irish-inflected Shakespearean performances was well established – and there was also evidence of the growth of that Irish tradition of staging Shakespeare that O'Toole had called for, an evolution intensified by the coincidence in the year 2016 of the fourth-hundredth anniversary of Shakespeare's death with the centenary of the Easter Rising against British rule in Ireland.

Open-air performances of *A Midsummer Night's Dream* have formed an identifiable stand within that play's production history since at least the nineteenth century, as Evelyn O'Malley explains in detail in *Weathering Shakespeare* (2020, pp. 45–94). There was evidence, however, of a further refinement in Parker's approach to the use of Irish accents in her production, making this *Dream* different from any that had been staged before. In the 2006 *Shrew*, the Irish accent was uniform: the actors all sounded as if they came from the same small Irish town or village. Contrastingly, in Kilkenny the cast spoke in a variety of Irish accents, displaying regional variation and thereby further undermining the belief that there is one 'correct' way to recite Shakespeare's verse in Ireland. That ethos was given force with the decision to cast a group of actors who in almost every case had recently graduated from Drama school, and who could thus be seen as representing the future of Irish theatre (one member of the cast, acting in one of his first professional roles, was Paul Mescal – who, by 2023, had become an Emmy- and Oscar-nominated actor). With a young cast and an unselfconscious approach to accent, this production represented a new and confident approach to staging Shakespeare in Ireland, providing some evidence of the emergence of that distinctive national tradition that O'Toole had bemoaned the absence of in 2012.

Parker's concern with *A Midsummer Night's Dream* was less with national politics than planetary catastrophe, however. The pre-publicity for her production stated that she would emphasize 'the crazy recklessness of mankind in the face of an ecosystem we understand imperfectly' (Rough Magic, 2018). It went on to describe the play as an attempt to address 'the dark sexuality of dreams, the supernatural, the mythic. Rough Magic will connect it to the here and now, a modern Ireland in the centre of global upheaval, a topsy-turvy universe, where the laws of physics have begun to warp.' One must avoid exaggerating the significance of what is, after all, advertising copy – but these statements capture well the foundational ideas for the production, as Parker herself made clear in press interviews. Speaking to the *Irish Times*, she discussed her 'fascination these days … with catastrophe on a human and global scale'. Referring to the speech by Titania that was discussed in the introduction to this Element, Parker describes those words as having 'more sweep and gravitas and majesty than anything else in the play':

> Of course [Shakespeare] weaves around it this fantastic farce of upended human emotions. But she says herself, 'This same progeny of evils comes from our dissent.' Their rage against each other is feeding into this dreadful upheaval in the climate …. Well, if you want a metaphor for what's going on right now, with G7, NATO and all that palaver. (Crawley, 2018)

Climate change, Parker stated, was not the 'be-all and end-all' of her approach to the production, but 'it's a really interesting part.' She also displayed an interest in Shakespeare's treatment of electricity as a form of 'natural magic'.

As Crawley explains, 'In her staging, electricity is another natural force in open revolt; once harnessed by humanity, like a beast of burden, it is now rebelling against us' (2018). In that directorial approach, one might find evidence to support Ghosh's ideas in *The Great Derangement*: Parker aims to reveal the truth of climate change but to do so she must draw on 'dreams, the supernatural, the mythic'; she must show how the laws of physics have begun to 'warp', while conveying the belief that natural processes are in 'rebellion'. This is Shakespeare as if re-imagined through the final act of Caryl Churchill's *Far Away* (2000); it is *A Midsummer Night's Dream* as 'akin to extraterrestrials or interplanetary travel'.

In presenting *A Midsummer Night's Dream* in this way, Parker was using the practice of theatrical revival to explore an idea that has frequently been considered in the scholarly literature: many studies of Shakespeare and ecocriticism refer to Titania's speech as an explication of how human behaviour might make the winds more destructive, might cause the rivers to overbear the continents. Parker made the contemporary relevance of Titania's speech tangible in many ways, perhaps the most obvious of which is the decision to stage the play outdoors, where it was at the mercy of the rain that is an inevitable feature of an Irish summer. As Sophie Chiari writes, in *A Midsummer Night's Dream*, 'wetness' informs the play as a whole (2018, p. 23), but the costume design by Katie Davenport (see Figure 3) reveals how an anticipated 'wetness' also affected the production: the actors' costumes were in many cases made of plastic or latex (materials that do not become excessively damp or sodden when wet), and the audience had been warned in advance that they would not be permitted to open umbrellas if (when) the rain fell.[3]

Those theatregoers also became more attuned to the relationship between time and their own bodily comfort due to the fact that the first act of the play coincided with the gradual setting of the sun (the starting time was 8 pm; sunset in Kilkenny in August occurs shortly after 9 pm). That timing also meant that the temperature was falling appreciably as the action continued into the fourth and fifth acts, resulting in a theatrical experience that was not so much outdoors as 'in the weather', to borrow an idea from O'Malley's *Weathering Shakespeare*. She argues in that book that outdoor performances can produce an awareness of 'environmental irony' (as, for instance, when a character refers to rainfall when an outdoor performance is presented under a cloudless sky), which can in turn reveal latent or unexamined environmental attitudes (p. 10). Thus, Hermia's line about how the 'tempest of my eyes' (her tears) could make up for a 'want of rain' in the forest setting of the play often

---

[3] My discussion is based on attendance on Saturday, 11 August 2018.

Figure 3 Titania in Rough Magic's *Midsummer Night's Dream*, played by
Martha Breen. Photo by Ste Murray

became unintentionally funny when delivered by an actor who was standing
under Irish rainfall (1.1. 130–31). These are examples of a heightened mode of
receiving Titania's words about what is now called climate change. Drawing
from the scholarship of Gwilym Jones (2016), O'Malley goes on to highlight
how such changes can be experienced as uncanny – an observation which one
might in turn link with Ghosh's 'great derangement' hypothesis, in that it
reveals how environmental encounters in the present can disrupt the correspondence between realistic representation of old plays and the lived experience of watching those plays in the present.

Also relevant was Parker's decision to stage the play in traverse: that is, the
action mostly unfolded between two banks of seats, each facing the other,
surrounded by the walls of the castle courtyard. Because of that setting, audience members were not only watching the action; they would also unavoidably
have witnessed the reactions of their fellow audience-members – an expression
and intensification of the communality of the experience. The two banks of seats

operated almost like two hemispheres of a globe – so that when Amy Conroy, the actor playing Puck, declared that she would 'put a girdle round about the earth/ In forty minutes' (2.1. 175–6), she enacted that idea by running in a circle behind the seating areas, doing a full circuit of the performance area in approximately one minute (rather than forty). In that moment, it was not the case that 'all the world's a stage' but rather that all the stage was representing the world (or 'the earth), and thus that the audience was a collective that might temporarily have represented all of humanity. Rather than giving rise to a weak feeling of the universality of human experience, it is likely that this strategy instead produced the kind of environmental irony that O'Malley writes about: the point was not that the audience represented 'the world' but rather that the audience was being made aware of its inability to do so adequately.

That limitation was caused partly by the fact that much of the 'earth' being represented in this production was human-constructed. As Figure 4 demonstrates, in the courtyard setting, the audience were largely cut off from

Figure 4 Kilkenny castle (photograph by Failte Ireland)

seeing their natural surroundings that are visible in the image; rather than were enclosed within the stone structure of walls and turrets that the image displays.

That courtyard did provide shelter from wind; it also enhanced the audibility of the performance (some of the actors' lines were amplified but most could be heard without the use of technology). It also had the potentially negative impact of removing natural referents (like the wooden pillars in the original Globe) that could be pointed at by the actors as they spoke of their surroundings. Thus, the actors could not gesture towards flowers, grass, or trees when speaking of such objects, but instead were required to use props or to rely on the audience's imagination. Yet in some ways, that approach was appropriate to the play: consider Theseus's observation in the final act that:

> the poet's eye, in fine frenzy rolling,
> Doth glance from heaven to earth, from earth to heaven;
> And as imagination bodies forth
> The forms of things unknown, the poet's pen
> Turns them to shapes and gives to airy nothing
> A local habitation and a name. (5.1. 14–18)

Again displaying Shakespeare's awareness that 'nothing' is never nothing, these lines show how the imagination can be used to conjure images of the natural world. But a point to be made about this production is that it may well be 'in the weather' but is not necessarily interacting with the natural world of non-human living beings. Instead, there was evidence of a desire to attune audiences' awareness to the differences between the real and the artificial, especially insofar as natural processes did (and did not) affect the audience.

Relatedly, the production also makes choices about gender and sexuality that highlight the dominance of the performative over the apparently real. One of the most common stereotypes associated with the Irish within English culture is of the Irish as inherently feminine – often in degradedly embodied and/or sexualized contexts. This is a trope that Shakespeare himself used in *The Comedy of Errors* when Ireland was described as being located 'between [the] buttocks' of a woman's body (3.2.124). Such embodiments of Ireland can be found in multiple sources from the sixteenth century onwards: in poetry, song, fiction, drama and (more recently) in cinema too. Put simply, Ireland was often characterized as a woman and England as a man – and the working out of their relationship via a marriage plot was implicitly based on an understanding of Ireland and England as existing in an interdependently binary relationship that is analogous to the patriarchal relationship between male and female (hence the significance of the gendered distinction between Béarla and Gaeilge referenced in the previous section). The Rough Magic *Midsummer Night's Dream* addresses those stereotypes in several ways, most obviously in casting the female actor Amy Conroy as Puck, a role more often played by male-identifying actors. Cross-gender casting has become more common in international productions of Shakespeare since the beginning of the twenty-first century, but the Irish approach to gender in Shakespeare is distinctive, both historically and in the contemporary period. Dublin, for example, was the place where a female performer first played Hamlet professionally (or, to be more precise, the earliest known record of such a performance taking place is in Dublin). That was in 1741 at Smock Alley Theatre, when the role was played by the English actress Fanny Furnival – several decades before such an act was considered permissible in London (see Howard, 2007, p. 1).

But the relationship between nation and gender was explored in other ways. The production drew an association between male dysfunction and imperial possession in its unambiguously negative presentation of Theseus's efforts to woo Hippolyta 'with my sword' (1.1. 16), alongside a characterization of Oberon as one-dimensionally petty and infantile. Also newly relevant was Egeus's determination to 'beg the ancient privilege of Athens' by putting his daughter to death for her refusal to marry the man he has chosen for her – a proposal that Theseus seeks to moderate by proposing she instead be banished to a convent. In an Ireland still coming to terms with the forced incarceration of thousands of Irish women in Magdalene Laundries run by the Catholic church, such words seemed anything but funny. Those moments collapsed the long history of patriarchal dominance of Ireland through colonialism, while also offering a reminder of how independent Ireland had treated its own female citizens.

Those explorations of gender – added to Parker's interest in the Irish voice and accent – explain the decision to stage the play in the castle courtyard, rather than in a setting that might have relied more directly on natural surroundings. However, the consequence of using that location was that audiences understood that the action was being framed by the castle, thus situating the play in the colonial context that also explains the approach to gender and voice. Kilkenny Castle was first built in the twelfth century by Norman invaders, before passing into the hands of the Butlers of Ormonde – staying in that line (and undergoing several refurbishments and rebuildings) until 1967 when the twenty-fourth Earl of Ormonde sold the castle to the Irish state for a nominal sum. Some traditions associated with the castle have a positive legacy (for example, as W. S. Clarke outlines (1955, p. 12), the Dublin Smock Alley company were invited to perform there in the late seventeenth century), while others are self-evidently negative (such as the fact that the upkeep of the castle was for much of its existence funded by rents that were paid by the dispossessed Catholic tenants who lived on nearby estates). Kilkenny Castle thus must be seen as both a symbol and a beneficiary of the English colonization of Ireland – which means that its existence forms part of the historical context outlined at the beginning of this section.

In addition, its colonial status puts the setting of the performance into dialogue with features of the play itself, which, as several scholars have argued, is inflected by the growth of English imperialism (for a useful summary of those debates, see Loomba, p. 8). Most prominently, Oberon and Titania's 'discord' is occasioned by a dispute over the possession of a child from India, allowing for the tracking of a lineage between Irish colonialism and the gradual expansion of England into India (marked, of course, by the foundation of the East India Company in 1600, one year after the Globe first opened). And, by linking those themes to the play's treatment of climate, Parker returns us to Lewis and Maslin's argument (and to Yusoff's discussion of it) that there are ethical implications to the decision about when to mark the beginning of the Anthropocene period, implications that create a continuum between climate, imperialism, and theatrical representation.

I have argued that Rough Magic's *Midsummer Night's Dream* explores colonialism, climate change and broader ecological matters, and Irish identity past and present. In doing so, the production can be seen as a staging of the 'conscious conjoining of differently scaled chronologies' that Dipesh Chakrabarty writes about in his discussion of how the Anthropocene requires new modes of cognition, research, and analysis (2021, p. 27). Returning to the questions that I began this section with, this can be seen as a production that retrieves and makes visible the historical context from which *A Midsummer Night's Dream* emerged – contexts that include the colonization of Ireland and the not unrelated ecological crisis that the scarcity of timber (and the long-term cold weather that the Orbis Spike is one measure of) signified. An 800-year-old castle, a 400-year-old play, and a cast of actors in their early twenties – all came together to demonstrate the overlapping modes of being human that, as Chakrabarty writes, 'we know at an abstract level' (p. 43) but too rarely see in material form – while also demonstrating the pertinence of Yusoff's argument that 'the Anthropocene exhibits a *colonial geology*' (her emphasis, p. 104). In the final section, I want to explain how those tendencies became enmeshed with the rise of Irish national theatre in the late nineteenth century.

3 Ecognosis: *DruidGregory* by Lady Gregory

One of the most famous objects in Irish theatre history is not a prop, a book, a building, or a costume – but a living being: the Coole Park Autograph Tree. A 300-year-old copper beech tree, it bears on its trunk the carved initials of many of the participants in the Irish Dramatic Revival – starting with Yeats, who was invited to add his name to the tree's bark in 1898 by Lady Gregory, the custodian of the Coole Park estate in south-east Galway, in the west of Ireland. Yeats and Gregory would go on to work together for more than three decades, establishing the Abbey Theatre in 1904 to support the development of a national (and sometimes nationalist) theatre tradition that allowed distinctive forms of acting and playwriting to emerge. That theatre also played an important role in the movement towards the achievement of Irish independence in 1922, and continues to have a prominent if occasionally controversial presence in Irish social and cultural life.

The carving of Yeats's initials was followed by inscriptions by many other literary visitors to Gregory's home. Those included playwrights (George Bernard Shaw, Sean O'Casey, Douglas Hyde, John Millington Synge), actors (Sara Allgood, the Fay Brothers), novelists (Violet Martin), and many others (see Figure 5). Their initials are gradually fading now: some have become illegible and, as the image indicates, they would soon be covered in moss if not for the attentions of the Coole Park groundskeepers. Yet, for now, the tree remains a monument to the Irish Dramatic Revival, showing how it was a complex network that had Coole Park – and Lady Gregory – at its centre: a status borne out by the fact that the trunk is now surrounded by a metal barrier that prevents any further additions from being made (see Figure 6).

Figure 5 The Coole Park autograph tree. Among the initials visible here are those of George Bernard Shaw (GBS, 2) and Augusta Gregory (AG, 8)

Figure 6 The Coole Park autograph tree

To carve one's name into a tree is an act of faith in a future that surpasses the span of a single human life. If left undisturbed, a copper beach tree can live for half a millennium, which means that the Coole Autograph Tree is at this time of writing heading into its late middle age. Yeats's initials will still be there long after most readers of this Element are gone – but Yeats's initials will probably be gone before the tree's life ends. To take a knife to the bark of a living tree is to see other living things, the so-called natural world, as a tabula rasa upon which humans may assert our identity. And yet the tree continues to grow, and the names continue to fade.

Coole Park is a place full of symbols. Visitors can often be seen counting swans, wondering if, as in Yeats's poem 'the Wild Swans at Coole', there are 'five and twenty' of them, with the odd number symbolizing the poet's loneliness. Lady Gregory's home – a place, Yeats wrote in 'Upon a House Shaken by the Land Agitation', 'where passion and precision have been one' – is no longer there, but its plinth remains, its absence monumental in its own way. To see a swooping bird in July in Coole is to be reminded of Yeats's remembrance of Synge and Hugh Lane in 'Coole Park, 1929': 'They came like swallows and like swallows went'. And although squirrels are much scarcer now than they were in Yeats and Gregory's time, one might still catch a glimpse of one and wonder why it runs away as if in mortal danger – as happens in 'To A Squirrel at Kyle-na-No'.

But the Autograph Tree will eventually prove resistant to any symbolism that is imposed upon it now: it will refuse to be encapsulated in solely human terms. The tree was there in 1768 when a man called Robert Gregory purchased the Coole estate, doing so with earnings from the East India Company (showing a further link between the colonization of Ireland and the later growth of British imperialism in Asia). The tree was also alive in 1847 when the name of William Gregory (later Augusta's husband) became 'one of the most detested in Ireland' when he added the Gregory Clause to the Poor Laws, thereby exacerbating the misery of many of the victims of the Great Irish Famine (Jenkins, 1986, p. 23). The tree was there when Coole Park passed from the Gregory family into the ownership of the newly independent Irish state – and it remains living today. So to see it only for its place in the Irish Revival – a movement that lasted from roughly 1890 to 1940 – is to shrink to human dimensions a living being that has borne witness to many phases of history: phases that have roots that spread to south Asia, north America, and elsewhere.

In one of the foundational texts of the environmental movement, *A Sand County Almanac* (1949), Aldo Leopold exhorted his readers to learn to think like mountains. Ireland does not have many mountains – just three peaks on the island are taller than 1,000 metres – so perhaps an Irish person might consider instead what it might mean to think like the Coole Park Autograph Tree. Doing so requires an understanding that human actions can remain evident in the skin of other living beings for hundreds of years after the death of the person who made the cut. And doing so allows one to see that, in the lifetime of our planet, human histories – of colonial appropriation, of famine, of national revival – are, to use a Yeats phrase from the suggestively titled 'To Ireland in the Coming Times', but the 'winking of an eye'.

To think like a mountain – or a tree – is to begin working with Timothy Morton's concept of 'ecognosis', a mode of ecological consciousness that is 'like knowing, but more like letting be known. It is something like coexisting. It is like becoming accustomed to something strange, yet also becoming accus-tomed to strangeness that doesn't become less strange through acclimation' (2016, p. 11). Morton further argues that one of the challenges presented by increased awareness of the link between humanity and rest of the planet is that '[w]e are faced with the task of thinking at temporal and spatial scales that are unfamiliar, even monstrously gigantic' (15) – an observation that applies to the discussion of deep time referenced briefly in the introductory section. The Anthropocene era thus requires an understanding of the agency of the individual in the context of a geological epoch that will persist for thousands of generations, and which seems likely to outlast the human species itself.

How can theatre give rise to 'ecognosis', both in its makers and its audiences? Can it produce encounters that involve becoming 'accustomed to strangeness' without that strangeness diminishing? Can it produce an awareness of the non-human natural world as co-existent with the human, rather than as a backdrop, symbol, or context for the human? To answer these questions, I explore *DruidGregory*, a production by Druid Theatre of five of Lady Gregory's plays, which were presented outdoors in Coole Park in September 2020 before touring to other rural locations in Galway in October of that year.[4] My intention is to show how Gregory revealed her detailed knowledge of the ecological: in her knowledge of agriculture, her spatial dramaturgy, her awareness of animals, and in other respects. By bringing the plays back to the place that inspired them, Druid revitalized that knowledge, allowing audiences to encounter it anew and to learn from it. Thus, the practice of revival can be used to refer not just to theatre practice but to ecological knowledge: under the direction of its Artistic Director Garry Hynes, Druid achieve both forms of revival simultaneously.

Druid Theatre was founded in 1975 in Galway, in the west of Ireland. It soon developed a reputation for carrying out what the company dubbed 'Unusual Rural Tours' – or 'URTs' for short. That involved staging plays in small towns across Ireland, usually in community halls and other improvised venues, and often in locations that had not previously had opportunities to host professional theatre. Those URTs had many impacts but collectively worked to challenge the spatial or place-based hierarchies that dominate Irish theatre. The national theatre movement of Yeats and Gregory had asserted that Irishness must be performed not in London but in Dublin – but to do that, they brought the indigenous cultures of the island from rural settings (such as Galway and the Aran Islands) to the urban centre (Dublin), thereby creating a new dynamic between core and periphery that drew accusations of inauthenticity: '*that's not the West* [of Ireland]' is what the protestors cried at the premiere of *The Playboy of the Western World* in Dublin in 1907. Staging Irish plays in Coole Park thus has symbolic value, connecting the artworks with the places that inspired their

---

[4] Comments on the production are based on my attendance at Coole Park (15 September 2020) and Kylemore Abbey (2 October 2020).

Figure 7 Publicity material for a production of *Cathleen Ni Houlihan* at the Abbey Theatre, 1984. Image reproduced by courtesy of the Abbey Theatre Archive. Note that only Yeats's authorship of the play is mentioned

creation in a way that breaks down distinctions between metropolitan centre and rural backwater.

Their Coole Park production was called *DruidGregory*, following other productions that involved the creation of compound words – *DruidSynge* (2005), *DruidMurphy* (2012), and *DruidShakespeare* (2015) among them. Some commentators see those titles, not unfairly, as an example of canny branding – but one might also argue that they capture well the ethos of Druid, and in particular the directorial style of Garry Hynes, which has often been evident in her interest in the techniques of dialogue and juxtaposition. That includes the dialogue that can arise between a play and the place that inspired it (as happened when Druid brought Synge to Aran); the dialogue between Irish and English culture that was captured by the *DruidShakespeare* name; and the juxtaposition of old plays with new contexts, as seen when Hynes revived Tom Murphy's 1966 play *Famine* in 2012, shortly after Ireland experienced an economic crash – allowing her to draw a link between the ecological catastrophe of the 1840s and Ireland's disastrous seduction by neoliberalism in the early twenty-first century.

Unlike those earlier cycles, *DruidGregory* was also intended as an act of canonization – placing Gregory where she belongs, alongside Beckett and Synge, as one of Ireland's most esteemed dramatists. She wrote more than thirty plays for the Abbey Theatre between 1904 and her death in 1932; she also gathered folklore (much of it taken from the Irish-speaking peasantry who lived on the Coole estate), wrote memoirs, and managed the Abbey with supreme effectiveness for almost thirty years. Yet her reputation is history is unmatched by her presence in the repertoire, from which she disappeared soon after her death. Indeed, in the fifty-year period between 1970 and 2020, the Abbey revived only three of Gregory's solo-authored plays: *The Gaol Gate* in 1971, *Coats* in 1973, and *The Workhouse Ward* in 1983. Those productions often appeared half-hearted: one was a lunchtime production, and the others appeared on double-bills alongside better-known plays. *Kathleen ni Houlihan* (1902), the play that (we now know) Gregory co-authored with Yeats, was given a full production twice at the Abbey during the same period, in 1984 and 1990 – though as was the practice at that time only Yeats's authorship of the play was mentioned in the publicity material (a point I will return to in more detail below, and see Figure 7).

During that half-century, however, Gregory was transformed within the Irish theatre from author to icon: she became a character in plays, novels, and films written by others, including John Ford, Carolyn Swift, Críostóir Ó Floinn, and Colm Tóibín – which is to say that the Abbey has presented more plays *about* Lady Gregory than *by* Lady Gregory since 1970. Her portrait has been displayed prominently in the Abbey's foyer throughout that period too, which means that – to borrow from the title of Cathy Leeney's pioneering 2002 anthology of plays by Irish women – she has become 'seen but not heard'.

DruidGregory began the work of setting that to rights. When performed at Coole Park, the production required audiences to move through the estate on foot, often following performers dressed in black who represented Gregory at different stages in her life. Chief among those actors was the Druid theatre co-founder Marie Mullen, whose resemblance to Lady Gregory was so close that a colourized photograph of Gregory was mistakenly captioned as Marie Mullen in at least three publications during the production's run. The casting of Mullen in that role had the impact of forming links with her other performances for Druid since 1975: she has played Pegeen Mike and the Widow Quin in Druid productions of Synge's *Playboy* in 1975 and 2005; played Mary in Tom Murphy's *Bailegngaire* at its premiere in 1985 and *Mommo* in the same play 30 years later; created the role of Maureen in McDonagh's *Beauty Queen of Leenane* in 1996 (winning a Tony Award when it transferred to Broadway) before taking on the role of Mag in that same play 30 years later. It is a truism that the body of an actor is a living archive of the roles they have played, but in bringing Mullen into correspondence with Gregory, an act of historiographical embodiment was taking place – rooting Gregory in a tradition of representation of Irish women that takes in Synge, Murphy, McDonagh and many others. This provides a physical manifestation of the 'matrilineal pattern' in Irish theatre

Figure 8 The plinth of Coole House, with illuminated window frames
highlighting the absence of the real building (photo by Matthew Harrison)

history that Melissa Sihra described as stretching from Gregory to Marina Carr
(2018, p. 11).

The *DruidGregory* performance begins with the recitation of one of
Gregory's poems on a pathway leading to her house. The design by Francis
O'Connor and Barry O'Brien creates a visual reminder of the house (which had
been demolished after Gregory's death), using lighting to signify window
frames (see Figure 8) – and they also included a desk with lit candles on it on
the plinth where the house used to stand. This clarified the link between
theatricality and the real in a way that persisted throughout the production:
the audience were not seeing an illusion that the house still existed but, on the
contrary, the use of theatrical illusion showed that it no longer exists. It was
difficult to conflate the real with the human in this setting; the audience thus
finds the natural setting being used (in the words of Morton) to allow them to
become 'accustomed to something strange'.

Audiences were then divided into smaller groups. The first play that I saw
was *The Gaol Gate* (1906), a drama set at the gates of Galway's municipal
prison, which in Gregory's time was situated where Galway Cathedral now
stands, at the heart of Galway city. The significance of its being staged in
Gregory's home is worth lingering upon. One of Gregory's occasional spats

with Synge arose when an Abbey company that was touring to Galway city staged her play there in 1908. 'I particularly didn't wish to have *Gaol Gate* [performed] there in the present state of agrarian excitement,' she wrote to her co-director; 'it would be looked on as a direct incentive to crime … '. But despite those objections, the play was staged for five nights. Gregory may have been more accurately expressing the truth when she then wrote that she 'could not ask "the classes" to come' to see her play in Galway because it was part of 'such a nationalist programme': she was willing to be a nationalist in Dublin but not in her home city, especially in front of the 'classes'. Her angst – and her ire – seemed to pass Synge by: in a letter to his fiancée Molly Allgood, he wrote that 'It is queer that Lady G. hasn't turned up'.

This is a funny moment in Irish theatre history but Gregory's anxieties, especially at that time in her life (and at that time in the political life of Ireland), were not just sincere but justified. Bringing the play back not just to Galway but to Coole thus represents a subtle reconciliation of Gregory's many allegiances, showing that her dual roles as Irish nationalist and pillar of the Anglo-Irish ascendancy could coalesce. To see the play being performed in Galway was thus to become aware of the passing of time, the passage of history, the lessening of the tensions between native and settler that have caused misery and death for much of the previous millennium.

The audience moved then to the edge of the Coole turlough for *The Rising of the Moon* (1907). A turlough is a body of water that is almost unique to the west of Ireland, with its name derived from the Irish words for 'dry' and 'lake' to signify the fact that the water disappears into a limestone base during the summer months before reappearing when rainfall increases later in the year. The apparent contradiction implicit in the Irish word indicates how that language offers a conception of the natural world that is different from what is possible in English: a turlough is a space that is *both* dry *and* a lake, and thus has an existence that spans time outside of what is immediately visible to the human subject.

But rather than looking at the turlough, the audience's gaze was instead directed to an illuminated reproduction of the moon, the natural object that gives the play its name. In another example of how the design sought to make the audience 'accustomed to something strange', the moon was represented by a self-evidently artificial stage property: yet again, what was being conveyed was the fact that theatre can display a version of the natural world but cannot replace it. Figure 9 shows the same prop later in the tour at Kylemore Abbey in Connemara, in north Galway. Again, the contrast between the living environment and the set-piece is very striking: the moon does not *represent* the natural world; if anything, it distracts from it.

Figure 9 The *Rising of the Moon* at Kylemore Abbey. The 'moon' is represented by the illuminated circle close the centre of the image (photo by Emilija Jefremova)

The juxtaposition of the theatrical and the natural persisted into later plays in the cycle. For example, the set for *Hyacinth Halvey* (1906) used a ruined building (formerly the estate's stables) as its own backdrop, showing how human constructions dwindle while the trees continue to grow (see Figure 10). The shift to the comedy *Hyacinth Halvey* (the third play that I saw) coincided with the movement of the environment from twilight into to darkness, allowing for more lighting to be used in the performance. The tones used to illuminate the stage mirrored the purples, pinks, and whites of the hyacinth flower, further delineating the real from the theatrical.

As the audience moved together towards the space for the performance of the penultimate play, I found myself increasingly aware of my own interconnection with the rest of the natural world. As night fell, I became cold, and I was also sometimes distracted from the drama by other living creatures. The action was now accompanied by the swooping of birds, which were feeding on insects that had emerged as the sunlight began to fade. Some insects had also been attracted by the presence of theatre lighting – which also meant that there were more birds than usual. The space was busy with life.

But now I wonder: is 'distraction' the right word? To use it implies that the birds and insects were not part of the stage spectacle, that my attention had gone somewhere that it did not belong. Yet the play being performed at that moment,

Figure 10 *Hyacinth Halvey* at Coole Park. The structure in the background is a ruin of the estate's stables, with trees visible behind it. The sheep carcass is fake. Photo by Matthew Harrison

McDonough's Wife (1912), is about what happens *elsewhere*. It is a strange play, and was performed only twice at the Abbey during Gregory's life (and never again thereafter) – yet it adopts a porous approach to space and character in ways that (as I discuss below) are a noteworthy characteristic of Gregory's dramaturgy overall, as shown by the fact that the eponymous wife never appears but is instead spoken about and, ultimately, becomes the subject of an improvised musical composition by her husband. Words push into music in this play: the natural into the supernatural, the scripted into the improvised, and the poor room in the family's house blends into a fair that is being held outside. Even the title seems unmoored: on stage at the Abbey, it was called *Macdara's wife* but in Gregory's collected plays the name changed to *McDonough's Wife*.

Given that the play seems to spill out in multiple directions, breaking down boundaries between what is on stage and what is not, it seems appropriate to imagine that it should be seen as including those birds and insects. Starting to think about other living beings as part of the performance means thinking about how theatre teaches audiences to see things – and not to see things. 'In a world certain humans have put under threat,' writes Evelyn O'Malley, 'the risks of not paying attention to all forms of performance in relation to their more-than-human counterparts are too high to ignore' (2020, p. 2); hence, the designation of other living beings as a 'distraction' in an era of mass extinction must be seen as more than a mere act of concentration on the performance of an artwork. O'Malley's statement also offers an example of how *DruidGregory* gave rise to ecognosis: the plays could have been presented in an outdoor setting (like Rough Magic's *Midsummer* in the courtyard) without bringing the audiences into contact with trees, grass, birds, and insects – but the choice to locate them in those potentially 'distracting' contexts had the impact of deepening and extending their meanings.

The staging choices for the first four plays in the cycle acted as a commentary on the conventions of drama from the Irish Revival. Many of the most famous realistic Irish plays of the Revival were enacted not in ecosystems but in rooms, with non-human natural phenomena appearing as context, decoration, or symbol – but rarely in and of themselves. They also often appear offstage: for 'culture' and 'nature', one could often insert 'indoors' and 'outdoors' respectively. This also meant that, within the Irish theatre of the Revival, the national was often conflated not with the landscape but with the home. As Nicholas Grene writes, audiences were often presented with 'A room within a house, a family within a room [which] stand in for nationality, for ordinary, familiar life; into the room there enters a stranger, and the incursion of that extrinsic, extraordinary figure alters, potentially transforms the scene' (1999, p. 52). In plays by writers such as Synge, the exterior world was often represented through windows and doors that, the audience must imagine, point not to the backstage area (as is actually the case) but to mountains, fields, the sea, or the sky. Audiences do not see the beach that Christy Mahon races on in *Playboy of the Western World*; they do not view the oceans that drown Maurya's sons in *Riders to the Sea*. Those entities are left offstage, literally waiting in the wings. The stage is where 'ordinary, familiar life' resides: the room is where it happens. Thus, when Irish dramatists of the Revival were attempting to shape their audiences' understanding of the real world as it appears within the theatrical frame, they were also shaping those audiences' understanding of the relationship between human and non-human beings.

An exception to this generalization is, however, the theatre of Lady Gregory. The staging of her plays outdoors by Druid had been necessitated by the COVID-19 pandemic (the Irish government had at that time banned indoor gatherings of more than five people). Seeing the plays in Coole Park provided a reminder that Gregory was unusual amongst the early Abbey dramatists in so often setting her plays outdoors, rather than in rooms: to return to Grene's phrase, for Gregory, 'ordinary, familiar life' was rarely to be found in a kitchen. One of the best examples of this tendency is her comedy *Spreading the News*, one of the three plays that opened the Abbey Theatre in 1904. As can

Figure 11 Set for *Spreading the News*, Abbey Theatre, 1904. Image reproduced by couresty of the Abbey Theatre Archive.

be seen from the original stage setting (see Figure 11), for Gregory the 'real world' was a natural environment that bore the trace of human intervention – the stone walls that break up the flow of the action, the painted backdrop that represents a continuum between the stage scene and the rest of the world.

Gregory used the insider/outsider dynamic to great effect in her dramaturgy and scenography thereafter. *The Gaol Gate* is about two women left on the threshold – neither in the gaol nor fully outside of it. The script of her play *The Bogey Man* (1912) was changed at the Abbey so that it was set on the edge of a town rather than 'where the coach stops' (as the published stage directions state). And her political play *The Wrens* (1914) was set not set in the eighteenth-century Irish parliament (another 'room where it happens') but on the street outside it. Gregory's plays – the five re-staged by Druid, and many others from her long career – thus resituate Irish life from the home to the external world. The revival of those plays can therefore offer an old form of dramaturgy as if it is new: even in the contemporary period, few Irish dramatists conceive of space in the way that Gregory did.

Gregory's plays also reveal a depth of knowledge of natural processes that is rarely found in contemporary Irish literature – which means that re-staging Gregory can allow for the retrieval of her knowledge. For example, to read or to watch her plays in the present is to be confronted with her detailed understanding of how food is produced. *Hyacinth Halvey*, to consider just one case, is a very funny play, but the presence on the stage of the dead body of a sheep shows Gregory's desire to contextualize the humour with her awareness that humans can live only by consuming something that has died. The original stage properties list from the Abbey archive shows Gregory's intentions: there is supposed to be a real sheep carcass on stage, not the illusion of one (as was used in *DruidGregory* – see Figure 10). Jessica Martell writes in relation to James Joyce that 'food does not merely constitute a semiotics in his work. It is also a material system whose figuration texturizes the somatic realities of colonial experience' in that food demonstrates how 'Irish modernism [is] the true literature of the Famine' (2020, p. 166). Something similar can be said of Gregory's dramaturgy which, if not obviously in the same tradition of modernism that Joyce occupied, nevertheless is determined by famine memory. Gregory in the present thus provides audiences with knowledge of the relationship between the human and the animal that they might otherwise be unaware of (or might prefer to ignore if they are aware of it).

The cycle, however, concluded with a play that *was* set indoors: *Kathleen ni Houlihan* – albeit that now the kitchen was represented outside, in a field behind the Coole Park visitor centre. Marie Mullen played the eponymous old woman, but in a way, she was almost playing Lady Gregory playing the character, something that had happened in 1919 when Gregory appeared in that role at the Abbey.

Thanks to the scholarship of James Pethica, the co-authorship of *Kathleen ni Houlihan* by Yeats and Gregory has been acknowledged since the late 1990s. One of the reasons that Gregory's co-authorship of the play is acknowledged is because she marked up a manuscript to show her contributions: 'All this mine alone' she wrote. That phrase could have been a subtitle for Druid's production of the play: Druid's *Kathleen ni Houlihan* had the impact not only of emphasizing Gregory's authorship of it but also of highlighting how it interconnects with the rest of her work. As curated by Druid, all five plays are shown to be dramas of the threshold – but they are also shown to dramas of transformation: the dead prisoner in *The Gaol Gate* becomes a martyr, and the policeman in *The Rising of the Moon* becomes a rebel, just as in *Kathleen ni Houlihan* the old woman becomes a 'young woman with the walk of a queen'. Characters in one play seem to recur in others – and not just because they are played by the same actors. Druid thus were showing that *Kathleen ni Houlihan* belongs in Gregory's *oeuvre* – that it might fit more comfortably there than in Yeats's collected plays.

The production thus underlined the value of intersectionality. Druid's staging of Gregory is an act of feminist theatre-making, an act of canon-re-formation – and thus it illustrates the extent to which climate justice overlaps with other forms of social justice. Gregory's exploration of indigeneity – through the Irish language and mythology – is also deeply relevant to our own times. The value of Druid's production was to show how these forms of knowledge – ecology, indigeneity, and gender – are interconnected, influencing and being influenced by each other: operating, in Julie Hudson's words, as 'feedback loops potentially reinforcing each other ... especially in the context of the broader cultural ecosystem' (p. 182) – and in turn illustrating the importance of Stacy Alaimo's designation of nature as a 'feminist space' (2019). In this Element, I have sought to highlight the extent to which the ecologies of Irish theatre draw on feminism, indigeneity, postcolonialism, and other epistemic frameworks; Druid's production shows those strands interacting and intersecting materially as well as conceptually.

If, like Yeats in 'the Wild Swans at Coole', we gaze upon those birds on the Coole turlough, we might recall his question about what might happen 'when I awake some day/ To find they have flown away'. The swans are likely to remain at Coole for the foreseeable future – but the swallows came to Galway later in 2020 than in earlier years, and they came in smaller numbers – so Yeats's question means something new now, and it is undeniably possible that his poetic imagery linking Synge and Hugh Lane (referred to above) with those birds may be less meaningful to future generations of readers, much as residents of Hampstead now listen in vain for the sounds of the nightingales that inspired Keats's ode. The revival of Gregory's plays by Druid shows how it is possible to retrieve the knowledge of figures like Yeats and Gregory – and to show how that knowledge speaks to those audiences' present and future.

The revitalization being dramatized by Druid was not just of Kathleen ni Houlihan, but of Gregory, of Coole Park, and of an Irish theatre that – because of the COVID-19 pandemic – had for months been under threat of collapse. Druid were not directly encouraging their audience to 'think like a tree', but their production was a remarkable feat of collapsing time, comparable to what Rough Magic did in making *A Midsummer Night's Dream* seem like a contemporary play – and to what Sarah Jane Scaife did in reconnecting Beckett to the indigenous Irish cultures that indirectly inspired his work. *DruidGregory* encouraged audiences to think at timescales that are more than human, even as the plays revealed the reverberations of human agency across several generations. It collapsed a century of theatre-making into a couple of hours, just as Coole Park and its Autograph Tree compress hundreds of years of human history into a single space, a single being. Both Gregory's play and Druid's production begin the task of allowing audiences to think about the intersection between the human and 'nature', inside and outside, past and present. Re-reading or re-watching Gregory – or, for many, reading or watching her for the first time – thus offered new ways of thinking like that tree – which is an Autograph Tree, for now, but also much more than that in the longer now that audiences everywhere must learn to understand.

Reflections

This Element began with the observation by the Anthropocene Working Group that the Anthropocene concept has developed differing meanings for different scholarly communities. What could it mean within theatre studies?

However one answers that question, I hope to have shown that theatre can certainly contribute knowledge to the debate. Bruno Latour has argued that one of the causes of ecological crisis is the disconnection of human subjectivity from the knowledge of the world that is engendered by the scientific method which, too often, sees itself as a 'the view from nowhere' and has thus become 'the new common sense' (2018, p. 68). Theatre is never a 'view from nowhere': it is a place-based (and thus an environmentally rooted) artform that creates communal experiences, and which also displays in material form the relationship between human decision-making and environmental impact. 'Thinking the Anthropocene,' write Bonneuil and Fressoz (2017), 'means challenging its unifying grand narrative of the errant human species and its redemption by science alone' (p. 288); in the discussion developed in this Element, my aim has been to show how theatre offers an effective method for challenging grand narratives and thus for allowing artists and audiences to think, feel, and live the Anthropocene.

Hence my choice of title: 'theatre revivals *for* the Anthropocene'. The history and practice of theatre can offer ways of understanding the world that are not so readily available within the sciences. Kathryn Yusoff has put forward the idea of 'Anthropocene monumentality' as 'a way to unpack the languages that geology carries and a way to push the conversation that admonishes the idea of the neutrality of geology as a language of the rocks and deep time, which is immune or innocent of its current deadly configurations' (p. 13). Theatre too can show how the human interaction with the ecological is never neutral and rarely disinterested – as evident in the present study from the fact that it seems impossible to explore performance ecology in Ireland without needing almost instantly to explore colonialism, patriarchy, and other formations and formulations of power.

Theatre also offers ways of imagining communal action, creating a meaningful space between the lone individual and the species in its entirety. As noted earlier, it is essential to take seriously the accusation that the Anthropocene paradigm blurs responsibility for environmental change by appearing to universalize human agency in a deeply unequal world. But against that consideration is the need to consider the risk that ecological crisis can further divide the world into separate camps, producing what Christian Parenti has (rather depressingly) called 'the politics of the armed lifeboat' (2011, p. 10). He imagines a future in which 'strong states with developed economies will succumb to a politics of xenophobia, racism, police repression, surveillance, and militarism and thus transform themselves into fortress societies while the rest of the world slips into collapse' (p. 20) – which means that the impacts of climate change will be unequally felt and unequally distributed throughout the world. The result, however – the extinction of the species – will by definition affect everyone. As Parenti puts it bluntly, 'no amount of walls, guns, barbed wire, armed aerial drones, or permanently deployed mercenaries will be able to save one half of the planet from the other' (p. 11).

This is a rather bleak prognostication but, in contemplating it, I am reminded of the value of Lynne Parker's decision to stage *A Midsummer Night's Dream* in traverse, thereby ensuring that every member of her audience was simultaneously watching the play while also watching their fellow audience-members' interpretation of the play. I do not wish to suggest naively that theatre can prevent rich countries from building walls to shut out climate refugees. But it can produce forms of community, empathy, and shared purpose that are otherwise in short supply. That is not a direct solution to the serious political or ecological problems that the human species faces. But it's also not nothing.

Theatre can also allow humans to understand the porousness of boundaries that they might otherwise consider absolute: not just the nature/culture divide that has so damagingly dominated Western culture for more than two centuries, but the barriers between the human and the non-human (thinking about the birds and insects that 'interrupted' *DruidGregory*), between past and present (the rendition of a Shakespeare play as if it was written for the Anthropocene), and between language and other forms of expression (as shown in the staging of *Laethanta Sona*). These three case studies were all performed outdoors (though *Laethanta Sona* also transferred to the Samuel Beckett Theatre in Dublin in October 2021), and thus made clear how the theatrical frame – the fourth wall – can be disturbed. How can other forms of theatre, including performances in theatre buildings, also teach audiences new ways of experiencing interconnection? There are many examples that could be considered in this context, in Ireland and elsewhere.

Theatre history can also demonstrate that human awareness of ecological interconnection is not new – but that it has at times been suppressed, distorted, and misunderstood. Lewis and Maslin have written about the need to prevent what they call 'the "accidental Anthropocene" story from taking root (p. 25)– the belief (popular amongst politicians, as they note) that humanity is not to blame for the climate crisis and thus should not be expected to make too demanding a sacrifice for the sake of future generations. Theatre from the past shows that the arrival of the Anthropocene was no 'accident'. The revival of old plays can force audiences now into a sense that urgent action is morally as well as scientifically essential.

That observation leads to a further question – is it possible to be hopeful?

Certainly, it is *difficult* to be hopeful, but it also necessary to acknowledge that ameliorative action is unlikely to happen without hope – and that hope can best be inspired by knowledge. Here I'm reminded of an anecdote told by the scientist Lawrence M. Krauss in the introduction to his 2021 book *The Physics of Climate Change*. That book's aim, he states, is to equip readers with information: about how climate change happens, why it is dangerous, and what can be done to stop its worst effects from happening. He uses simple physics in an accessible style ('it's not rocket science!' he quips), and makes complex ideas seem graspable. It is thus an absorbing but also a *hopeful* read. Yet, as Krauss reveals in that introduction, he found it exceedingly difficult to find a publisher for the book, despite having written several best-selling popular volumes in the past. The reason? 'Numerous publishers and editors indicated to me', he explains, that 'the only marketable books on climate change would be ones that appeal to emotions and communicate to true believers through a sense of doom and gloom' (2021, p. 5). His book's appearance, however, illustrates that knowledge can prevent a sense of doom – a sense of hopelessness or inaction (a sense which ultimately leaves individuals with no option but to continue *consuming* – through social media, by doom-streaming apocalyptic documentaries, and so on – instead of acting to produce change). Accurate general knowledge of the scientific basis of ecological change is essential but, as I hope the examples explored in this Element have shown, theatre can produce other forms of knowledge too, such as the different forms of *ecognosis* that are inspired by the three productions I have discussed. Scholars must do more than simply 'appealing to emotions' and it is surely relevant in that context that none of the productions explored in this Element produces 'doom and gloom' in relation to environmental matters but instead aims to provoke thought, know-ledge, and new ways of seeing. Knowledge and hope can also be 'missiles to the future', to return to Malm's suggestive phrasing.

As Latour writes, 'the Anthropocene, despite its name, is not an immoderate extension of anthropo*centrism*, as if we boast of having really been changed into Supermen of sorts . . . It is rather the human as a unified agent, as a simple virtual political entity . . . that has to be decomposed into several distinct peoples, endowed with contradictory interests, competing territories' (2018, p. 122). In exploring case studies from Ireland (one of 'several distinct peoples') I have sought to consider how those individual instances can allow for an analysis that might be applicable to 'the human as a unified agent'. Rather than stating that there are limitations to that approach (which is self-evident), I prefer to conclude by expressing a belief (probably also self-evident) that there is a great deal more to be said (and staged) in relation to these topics: the perspectives presented here are intended to stimulate comparison rather than being applied directly into other contexts; they are intended to contribute to a debate rather than offering conclusive findings.

Writing in 2018, Carl Lavery suggested that the ecological occupies a position of 'relative marginality' within 'the disciplines of theatre and performance studies' (p. 23). That statement was accurate at the time of writing but I am uncertain if it remains so now: a greater number of ecocritical publications are appearing, work from the 1990s and early 2000s is being cited with growing frequency, and perhaps one indication of the vitality of the topic within the field is the fact that the 2021 conference of the International Federation for Theatre Research, on the theme of theatre ecologies, featured more than 900 papers by scholars from more than 80 countries (that online conference was co-organized by me and colleagues at the University of Galway). Perhaps scholars in this field can aspire to bring about a future in which the *eco-* prefix becomes redundant – in which criticism, dramaturgy, sceneography, and the practice of making and reviving performances all contain an inherent apprehension of the importance of the ecological for the encounter between theatre-makers, audiences, and the rest of our shared world. 'All theatre,' writes Julie Hudson, 'can be, in some sense, ecotheatrical' (p. 5); it should be possible to hope for a future in which such statements are seen not as assertions to be argued for but as an expression of a first principle that is shared by makers and audiences alike.

This returns us to the idea of future fossils. It is apparent that the plays discussed in this volume will produce future fossils. That is literally the case in the sense that they make use of materials (steel and plastic, for instance) that will continue to exist long after their makers have died. But it is also necessary to think of theatrical performance now as productive of future remnants in other ways, as shown by my discussion of how the decisions made in the time of Shakespeare, Gregory, and Beckett have informed and shaped the present. The time that audiences find themselves in now encompasses the long now of the geological. But it must also include the long now that entails the impact of our actions upon future generations. Theatre is proverbially 'live' in the sense that it happens in a moment and is then, it is often said, gone forever. But it is now necessary to show that even the immediately live is never gone forever: there will be reverberations, traces, and consequences that must be known, and accounted for.

References

Ahmadi, Mohebat, 2022. *Towards an Ecocritical Theatre: Playing the Anthropocene*. London: Routledge.

2015. 'Andrew Bovell's "when the rain stops falling": Theatre in the age of "hyperobjects"'. *Australasian Drama Studies* 66, pp. 40–62.

Alaimo, Stacey, 2017. 'Anthropocene Feminism'. In *Anthropocene Feminism*, ed., Richard Grusin. Minneapolis: University of Minnesota Press, pp. 89–121.

2019. *Undomesticated Ground*. New York: Cornell University Press.

Angelaki, Vicky, 2019. *Theatre and Environment*. London: Bloomsbury.

Anthropocene Working Group, 2022. 'Introduction'. http://quaternary.stratigraphy.org/working-groups/anthropocene/ Accessed 21 March 2021.

Arbuthnot, Sharon, Ni Mhaonaigh, Maire, and Toner, Gregory 2019. *A History of Ireland in 100 Words*. Dublin: RIA.

Arons, Wendy, 2020. 'Tragedies of the capitalocene'. *Journal of Contemporary Drama in English* 8.1, pp. 16–33.

Arons, Wendy, and May, Theresa J., eds., 2012. *Readings in Performance and Ecology*. New York: Palgrave Macmillan.

Aston, Elaine, 2015. 'Caryl Churchill's "dark ecology"'. In *Rethinking the Theatre of the Absurd: Ecology, the Environment and the Greening of the Modern Stage*, eds. Lavery, Carl and Finburgh, Clare pp. 59–76. London: Bloomsbury Publishing.

Balestrini, Nassim W., 2020. 'Sounding the Arctic in Chantal Bilodeau's climate change plays'. *Nordic Theatre Studies* 32.1, pp. 66–81.

Beckett, Samuel, 2010a. *Complete Dramatic Works*. London: Faber.

2010b. *More Pricks than Kicks*. London: Faber.

1961. *Happy Days*. New York: Grove Press.

1963. *Oh les beaux jours*. Paris: Seuil.

2021. *Laethanta Sona*, translated by Micheál Ó Conghaile. Inverin: Cló Iar-Chonnacht.

Beer, Tanja, 2021. *Ecosceneography*. Chaim: Palgrave.

Behringer, Wolfgang, 2010. *A Cultural History of Climate*. Cambridge: Polity.

Ben-Zvi, Linda, ed., 1992. *Women in Beckett: Performance and Critical Perspectives*. Chicago: University of Illinois Press.

Bonneuil, Christophe, and Fressoz, Jean-Baptiste, 2017. *The Shock of the Anthropocene: The Earth, History and Us*. London: Verso.

Bould, Mark, 2021. *The Anthropocene Unconscious*. London: Verso.

Bruckner, Lynne D. and Brayton, Dan, eds., 2011. *Ecocritical Shakespeare.* Burlington: Ashgate.

Chakrabarty, Dipesh, 2021. *The Climate of History in a Planetary Age.* Chicago: University of Chicago Press.

Chaudhuri, Una and Enelow, Shonni, 2013. *Research Theatre, Climate Change, and the Ecocide Project: A Casebook.* New York: Palgrave .

Chaudhuri, Una, 1994. '"There must be a lot of fish in that lake": Toward an ecological theater'. *Theater* 25.1, pp. 23–31.

1997. *Staging Place: The Geography of Modern Drama.* Ann Arbor: University of Michigan Press.

Chiari, Sophie, 2018. *Shakespeare's Representation of Weather, Climate and Environment: The Early Modern 'Fated Sky'.* Edinburgh: Edinburgh University Press.

Clark, W. S., 1955. *The Early* Irish *Stage: The Beginnings to 1720.* Oxford University Press.

Cless, Downing, 2010. *Ecology and Environment in European Drama.* London: Routledge.

Collins, Lucy, 2021. '"Nature herself seems in the vapours now": Poetry and climate change in Ireland 1600–1820'. In *Climate and Society in Ireland,* eds., James Kelly and Tomás Ó Carragáin. Dublin: RIA, pp. 325–48.

Crawley, Peter, 2018. 'It's a kind of magic'. *Irish Times,* 4 August. www.irishtimes.com/culture/stage/it-s-a-kind-of-magic-electric-midsum mer-night-s-dreams-in-kilkenny-castle-1.3580898.

Croall, John, 2018. *Performing Hamlet: Actors in the Modern Age.* London: Bloomsbury.

Crutzen, Paul, 2002. 'The "Anthropocene"'. *Journal de Physique IV (Proceedings)* 12.10, pp. 1–5.

Crutzen, Paul and Steffen, Will, 2003. 'How long have we been in the Anthropocene era?' *Climactic Change,* 61, pp. 251–7.

Davis, Janae, Moulton, Alex A., Van Sant, Levi, and Williams, Brian, 2019. 'Anthropocene, capitalocene … plantationocene? A manifesto for ecological justice in an age of global crises'. *Geography Compass* 13.5, p. e12438.

Diamond, Elin, 2004. 'Feminist readings of Beckett'. In *Palgrave Advances in Samuel Beckett Studies,* ed., Lois Oppenheim. London: Palgrave Macmillan, pp. 45–67.

Dobson, Michael, 2011. *Shakespeare and Amateur Performance: A Cultural History.* Cambridge: Cambridge University Press.

Egan, Gabriel, 2006. *Green Shakespeare: From Ecopolitics to Ecocriticism.* London: Routledge.

Egan, Gabriel, 2015. *Shakespeare and ecocritical theory*. London: Bloomsbury Publishing.

Everett, Nigel, 2014. *The Woods of Ireland: A History, 700–1800*. Dublin: Four Courts Press.

Fakhrkonandeh, Alireza, 2021. 'Oil cultures, world drama and contemporaneity: Questions of time, space and form in Ella Hickson's oil'. *Textual Practice*, 36(1), pp. 1–37.

Farrier, David, 2019. *Anthropocene Poetics: Deep Time, Sacrifice Zones, and Extinction*. Minneapolis: University of Minnesota Press.

2021. *Footprints: In Search of Future Fossils*. London: Picador.

Fitzgerald, Lisa, 2017. *Re-Place: Irish Theatre Environments_* Oxford: Lang.

Flood, Alison, 2013. 'Dan Brown's Inferno heats up book sales'. *Guardian* 21 May. www.theguardian.com/books/2013/may/21/dan-brown-inferno-book-sales.

Fressoz, Jean-Baptiste and Locher, Fabien, 2020. *Les Révoltes du ciel: Une histoire du changement climatique XVe-XXe siècle*. Paris: Seuil.

Fressoz, Jean-Baptiste, 2012. *L'apocalypse joyeuse. Une histoire du risque technologique*. Paris: Seuil.

Friel, Brian, 2013. *Plays 1*. London: Faber.

Garrard, Greg, 2011. '"Endgame": Beckett's "ecological thought"'. *Samuel Beckett Today / Aujourd'hui* 23, pp. 383–97.

Ghosh, Amitav, 2016. *The Great Derangement: Climate Change and the Unthinkable*. Chicago: University of Chicago Press.

Giannichi, Gabriella and Stewart, Nigel, 2005. *Performing Nature*. Oxford: Lang.

Gilbert, Helen, 2020. 'Indigenous festivals in the Pacific: Cultural renewal, decolonization and nation-building'. *Pacific and American studies* 20, pp. 41–56.

2019. 'Performing the Anthropocene'. In *Ecocritical Concerns and the Australian Continent*, eds., Beate, Neumeier, and Helen, Tiffin. Lanham: Rowman and Littlefield, pp. 219–34.

2013a. 'Indigeneity, time and the cosmopolitics of postcolonial belonging in the atomic age'. *Interventions* 15, pp. 195–210.

2013b. 'Indigeneity and performance'. *Interventions* 15.2, pp.173–80.

Gillen, Katherine, 2018. 'Shakespeare in the Capitalocene: *Titus Andronicus, Timon of Athens*, and early modern eco-theater'. *Exemplaria* 30.4, pp. 275–92.

Grene, Nicholas, 1999. *The Politics of Irish Drama*. Cambridge: Cambridge University Press.

Grusin, Richard, ed., 2017. *Anthropocene Feminism*. Minneapolis: University of Minnesota Press.

Haraway, Donna, 2016. *Staying with the Trouble*. Durham: Duke University Press.

Heaney, Seamus, 1980. *Preoccupations*. London: Faber.

Herron, Thomas, 1998. 'Goodly woods': Irish forests, georgic trees in books 1 and 4 of Edmund Spenser's Færie Queene'. *Quidditas* 19.1, pp. 97–122.

Howard, Tony, 2007. *Women as Hamlet: Performance and Interpretation in Theatre, Film and Fiction*. Cambridge: Cambridge University Press.

Hudson, Julie, 2020. *The Environment on Stage: Scenery or Shapeshifter?* London: Routledge.

Jenkins, Brian, 1986. *Sir William Gregory of Coole*. Gerrards Cross: Colin Smythe.

Jones, Gwilym, 2016. *Shakespeare's Storms*. Manchester: Manchester University Press.

Joyce, James, 2000. *Ulysses*. Harmondsworth: Penguin.

Kershaw, Baz, 2007. *Theatre Ecology: Environments and Performance Events*. Cambridge: Cambridge University Press.

Kiberd, Declan, 2001. *Irish Classics*. Cambridge: Harvard University Press.

1995. *Inventing Ireland*. Cambridge: Harvard University Press.

Knowlson, James, 1997. *Damned to Fame*. London: Bloomsbury.

Krauss, Laurence M., 2021. *The Physics of Climate Change*. New York: Schuster.

Kronik, Jakob and Verner, Dorte, 2010. *Indigenous Peoples and Climate Change in Latin America and the Caribbean*. Washington DC: World Bank.

Latour, Bruno, 2017. *Facing Gaia: Eight Lectures on the New Climatic Regime*. Oxford: John Wiley.

2018. *Down to Earth: Politics in the New Climatic Regime*. Cambridge: Polity

Lavery, Carl, ed., 2019. *Performance and Ecology: What Can Theatre Do?* London: Routledge.

2018. 'Ecology in Beckett's theatre garden: Or how to cultivate the oikos'. *Contemporary Theatre Review*, 28:1, pp. 10–26.

Leeney, Cathy, 2002. *Seen and Heard*. Dublin: Carysfort.

Lewis, Simon and Maslin, Mark A., 2018. *The Human Planet: How We Created the Anthropocene*. London: Pelican.

2015. 'Defining the Anthropocene'. *Nature* 519, pp. 171–80.

Lonergan, Patrick, 2022. '"A missile to the future" – the theatre ecologies of Caryl Churchill's *Far Away* on Spike Island'. *Journal of Contemporary Drama in English*, 10(1), pp. 133–47.

2020. '"A twisted, looping form" staging dark ecologies in Ella Hickson's *Oil*'. *Performance Research* 25.2, pp. 38–44.

2019. *Irish Drama and Theatre since 1950*. London: Bloomsbury.

2015. 'Shakespearean productions at the Abbey Theatre, 1970–1985'. In *Irish Theatre in Transition*, ed., Donald Morse. London: Palgrave Macmillan, pp. 149–61.

Loomba, Ania, 2002. *Shakespeare, Race, and Colonialism*. Oxford: Oxford University Press.

Love, Catherine, 2020. 'From facts to feelings: The development of Katie Mitchell's Ecodramaturgy'. *Contemporary Theatre Review* 30.2, pp. 226–35.

MacFaul, Tom, 2015. *Shakespeare and the Natural World*. Cambridge: Cambridge University Press.

Malm, Andreas, 2016. *Fossil capital: The rise of steam power and the roots of global warming*. London: Verso.

Martell, Jessica, 2020. *Farm to Form*. Reno: University of Nevada Press.

Martin, Randall, 2015. *Shakespeare and Ecology*. Oxford: Oxford University Press.

May, Theresa J., 2005. 'Greening the theater: Taking ecocriticism from page to stage'. *Interdisciplinary Literary Studies* 7.1, pp. 84–103.

2021. *Earth Matters on Stage*. London: Routledge.

McAllister, Jonathan, 2022. 'Happy days, directed by Trevor Nunn'. *Journal of Beckett Studies*, 31(2), pp. 226–9.

McMullan, Anna, 2021. *Beckett's Intermedial Ecosystems: Closed Space Environments across the Stage, Prose and Media Works*. Cambridge: Cambridge University Press.

Mentz, Steve, 2015. *Shipwreck Modernity: Ecologies of Globalization, 1550–1719*. Minneapolis: University of Minnesota Press.

Moore, Jason W., ed., 2016. *Anthropocene or Capitalocene? Nature, History, and the Crisis of Capitalism*. Oakland: Pm Press.

Morash, Chris, 2002. *A History of Irish Theatre*. Cambridge: Cambridge University Press.

Morton, Timothy, 2016. *Dark Ecology: For a Logic of Future Coexistence*. New York: Columbia University Press.

2013. *Hyperobjects: Philosophy and Ecology after the End of the World*. Minneapolis: University of Minnesota Press.

Nardizzi, Vin, 2011. 'Shakespeare's globe and England's woods'. *Shakespeare Studies* 39, pp. 54–63, 16.

Ni Chongaile, Cathleen, 2021. 'Programme note'. *Beckett sa Chreig/Laethanta Sona show programme*.

Nixon, Rob, 2011. *Slow Violence and the Environmentalism of the Poor*. Cambridge: Harvard University Press.

O'Malley, Evelyn, 2020. *Weathering Shakespeare: Audiences and Open-air Performance*. London: Bloomsbury.

O'Toole, Fintan, 2012. 'Wherefore art thou, Irish rat?'. *Irish Times*. 25 August. Saturday Pg. 8. www.irishtimes.com/culture/tv-radio-web/wherefore-art-thou-irish-rat-1.543055

Parenti, Christian, 2011. *Tropic of Chaos*. New York: Bold Type.

Parham, J., ed., 2021. *The Cambridge Companion to Literature and the Anthropocene*. Cambridge: Cambridge University Press.

Prateek, 2020. 'The politics of dark ecologies in Deepan Sivaraman's Peer Gynt'. *Performance Research* 25.2, pp. 134–40.

Read, Alan, 1995. *Theatre and Everyday Life: An Ethics of Performance*. London: Taylor & Francis.

Rough Magic, 2018. '*A midsummer night's dream*'. www.roughmagic.ie/archive/a-midsummer-nights-dream/.

Scaife, Sarah, Jane, 2016. 'Practice in focus: Beckett in the city'. *Staging Beckett in Ireland and Northern Ireland*. London: Bloomsbury, pp. 153–67.

Shapiro, James, 2006. *1599*. London: Faber.

Shaw, Fiona, 2008. 'Many happy days'. *New England Review (1990-)* 29.4, pp. 111–3.

Sihra, Melissa, 2018. *Marina Carr: Pastures of the Unknown*. Cham: Springer.

Steffen, Will, Broadgate, Wendy, Deutsch, Lisa, Gaffney, Owen, and Cornelia Ludwig 2015. 'The trajectory of the Anthropocene: The great acceleration'. *The Anthropocene Review* 2.1, pp. 81–98.

Steffen, William H (2023), *Anthropocene Theater and the Shakespearean Stage_*.Oxford: Oxford University Press.

Stevens, Lara, Tait, Peta, and Varney, Denise, eds., 2017. *Feminist Ecologies: Changing Environments in the Anthropocene*. Cham: Springer.

Streeby, Shelley, 2018. *Imagining the Future of Climate Change: World-Making through Science Fiction and Activism*. Berkeley: University of California Press.

Sutoris, Peter, 2021. 'Decolonise the Anthropocene'. https://theconversation.com/the-term-anthropocene-isnt-perfect-but-it-shows-us-the-scale-of-the-environmental-crisis-weve-caused-169301.

Synge, J. M., 1982. *Collected Works, Volume 3: Plays 1*, ed. Ann Saddlemyer. Gerrards Cross: Colin Smythe. www.amazon.co.uk/J-M-Synge-Collected-Works-Plays/dp/B003079SD0

Synge, J. M., 1983. *Collected Works 2: Prose*, ed. Alan Price. Gerrards Cross: Colin Smythe. www.amazon.co.uk/J-M-Synge-Collected-Works-Plays/dp/B003079SD0

Thomas, Claire, 2020. *The Performance*. London: Weidenfeld & Nicolson.

Varney, Denise, 2022. 'Caught in the Anthropocene: Theatres of trees, place and politics'. *Theatre Research International* 47.1, pp. 7–27.

Walsh, John, 2022. *One Hundred Years of Irish Language Policy, 1922–2022.* Oxford: Lang.

Wark, McKenzie, 2015. *Molecular red: Theory for the Anthropocene.* Verso Books.

Wilmer, Steve and Vedel, Karen, 2020. 'Theatre and the Anthropocene: Introduction'. *Nordic Theatre Studies*, 32.1, pp. 1–5.

Woynarski, Lisa, 2020. *Ecodramaturgies.* Cham: Springer.

Yusoff, Kathryn, 2018. *A Billion Black Anthropocenes or None.* Minneapolis: University of Minnesota Press.

Acknowledgments

I wish to acknowledge with thanks the permission of Company SJ, Rough Magic, Ste Murray, Druid Theatre, and the Abbey Theatre to reproduce images. Parts of this Element were developed through lectures, conference papers, and classes – and I am grateful to the many people who offered invitations to speak, provided feedback, or offered other forms of support. Thanks especially to Vicky Angelaki, Ina Bergmann, Ashley Cahilane, Maria Eisenmann, Nicholas Grene, Miriam Haughton, Yeeyon Im, Marianne Kennedy, Michal Lachman, Charlotte McIvor, Martin Middeke, Laoighseach Ni Choistealbha, Martin Riedelsheimer, Malcom Sen, Kurt Taroff, Adam Versenyi, and Ian Walsh. I also want to thank the members of the IFTR Theatre Historiography Working Group, who offered feedback on a draft of Section 3, and the organizers and participants of the 'Shakespeare and Climate Emergency Conference' at the Globe Theatre in April 2021, at which I delivered an early draft of parts of Section 2. I benefitted enormously from the advice of two anonymous readers, and I thank them for their judicious and constructive feedback. My thanks also to the series editors, Trish Reid and Liz Tomlin. Finally, my thanks and love as always to Therese, Saoirse, and Cónall.

Cambridge Elements ☰

Theatre, Performance and the Political

Trish Reid
University of Reading

Trish Reid is Professor of Theatre and Performance and Head of the School of Arts and Communication Design at the University of Reading. She is the author of *The Theatre of Anthony Neilson* (2017), *Theatre & Scotland* (2013), *Theatre and Performance in Contemporary Scotland* (2024) and co-editor of the *Routledge Companion to Twentieth-Century British Theatre* (2024).

Liz Tomlin
University of Glasgow

Liz Tomlin is Professor of Theatre and Performance at the University of Glasgow. Monographs include *Acts and Apparitions: Discourses on the Real in Performance Practice and Theory* (2013) and *Political Dramaturgies and Theatre Spectatorship: Provocations for Change* (2019). She edited *British Theatre Companies 1995 - 2014* (2015) and was the writer and co-director with Point Blank Theatre from 1999-2009.

About the Series

Elements in Theatre, Performance and the Political showcases ground-breaking research that responds urgently and critically to the defining political concerns, and approaches, of our time. International in scope, the series engages with diverse performance histories and intellectual traditions, contesting established histories and providing new critical perspectives.

Cambridge Elements ≡

Theatre, Performance and the Political

Elements in the Series

Theatre, Performance and the Political
Patrick Lonergan

A full series listing is available at: www.cambridge.org/ETPP

Printed in the United States
by Baker & Taylor Publisher Services